stargazing
secrets

ANTON VAMPLEW

FELLOW OF THE ROYAL ASTRONOMICAL SOCIETY

Collins

For Gillian

HarperCollins Publishers Ltd.
77-85 Fulham Palace Road
London
W6 8JB

The Collins website address is:
www.collins.co.uk

Collins is a registered trademark of
HarperCollins Publishers Ltd.

First published in 2007

12 11 10 09 08 07
10 9 8 7 6 5 4 3 2 1

A catalogue record for this book is available from the British Library.

ISBN 978 0 00 724224 5

Collins uses papers that are natural, renewable and recyclable
products made from wood grown in sustainable forests. The
manufacturing processes conform to the environmental regulations
of the country of origin.

Cover design by Emma Jern
Design layout by Richard Marston
Design by D & N Publishing
Edited by Caroline Taggart

Printed and bound by Printing Express Ltd., Hong Kong

Contents

Introduction

Prepare yourself for a further adventure...

...full of fast-moving science in a Universe that gently meanders along with time, regardless of what we're up to. We live in a truly exciting astronomical time, but what future historians will make of our energetic efforts to unravel the secrets of the stars is anybody's guess. They may smile at our ideas as we smile at some of those of previous generations – like the people who believed the Universe was on the back of an extremely large (and I mean extremely large) turtle. Is that crazy? I quite like the idea that there are four sturdy feet holding us 'up'. All the early ideas went hand-in-hand with the technology of the time and the thinkers who came before. Today we have a great knowledge base built up over several thousand years, incredible telescopes and space probes for capturing information, and supercomputers to analyse (as well as they've been programmed to) all that can be thrown at them.

Away from the mainstream of starry discovery, a tributary of astronomers has been looking for mysterious dark stuff and ghostly particles that bind the Universe by weaving through higher dimensions. Cool or what? This dark stuff may or may not exist and may or may not have mass which, they say (and I am always dubious about those who call themselves 'they'), will determine whether the Universe expands forever – or doesn't – and whether it will eventually collapse back into what I imagine will be a splendid mess at the end of the show. 'They' have the entire fate of the Universe down to a single number – and no, they have no idea what the number is. That's not so cool, is it?

We know very little about our 'moment' of cosmic history in this respect. Where do we fit in along the whole timeline of the Universe? Near the beginning? Middle-ish? Didn't the Mayan calendar say we have just a few years left? I'd love to know if the aliens from that nice warm watery planet around epsilon Eridani have the same thoughts. They must know that one day their star will have exhausted all its fuel and then they'll need to find a new world – won't they? How inquisitive are aliens anyway? We definitely are, or we wouldn't have gone to the Moon or be exploring the outer reaches of the solar system in order to work out how planets and stars are made, or trying to figure out where and how we came into existence to ask these questions.

You now ask: 'Has this got anything to do with finding the Plough or the Southern Cross?' Well, while I may not actually give a clear answer to that, I do believe that space really is the biggest subject ever – one moment you're trying to find the star alpha Equulei in the sky and the next you're steering a course down the starry river of 'I wonder?' You really can wade in as deeply or as shallowly as you wish – just enjoy it.

Starry terms you find in bold are defined in the Astroglossary at the back.

CHAPTER 1

Starry Skies from a Spinning Planet

Take a look in the mirror. Now tell me the Earth has nothing to answer for. What a planet!

Indeed, what an amazing species the Earth has allowed us to become. The very nature of our orbit around the Sun, which happens to lie within the right zone – not too far and not too close – has, because we are situated precisely where we are, encouraged our sort of life. What would we be like if the Earth had been slightly closer to or further from the Sun? Or spun faster or slower? I mean it takes 243 Earth days for Venus to spin once – that's how long one Venusian day is. Would we have evolved as 'us' under these circumstances? The Earth has allowed me to ask this question. But I could just as easily have asked, 'What's for tea?' How great are we?! We can happily flip from deep questions of existence to everyday ones in a moment.

Now, you may have informed ideas about life starting on a slightly different Earth, but ultimately there is no right or wrong answer.

It's just like the 'Do aliens exist?' question: no one knows for sure, so no one can be proved right or wrong, whatever arguments are put forward. Other questions can be answered with pure descriptions of what happens, but even these can still be pretty tricky. I always like 'Why does the Moon look big close to the horizon, but smaller when higher up?', as it needs a bit of a run-up before you splat the answer in the questioner's face. For the first time, in Chapter 9, I have tried to describe this phenomenon as clearly as possible, and it demonstrates perfectly the cunning nature of some seemingly innocent questions.

Let's have a look at the Earth for some more examples of 'Well, that's easy for you to ask' questions that need some serious attention when answering. Here's one: why do we see different constellations at different times of year? Let's start the answer with another simple question: how long does it take the Earth to spin around once?

Twenty-four hours? We all know that the Earth spins once a day. Aha! Intriguingly there are *two* different kinds of day (actually there are several others, but they are really only for the serious wobble-and-tilt enthusiast). Why is this? It is because as well as the Earth spinning on its axis it is also orbiting the Sun. Which leads to two days of *different* lengths of time. Oh.

Our 24-hour-clock time is based on where the Sun is, and this is called the **solar day** (also known as a **synodic** day – see the Astroglossary at the back of the book). However, the Earth actually spins around on its axis exactly once in a shorter 23 hours 56 minutes and 4 seconds – this is called the

sidereal day, which is the time relative to the 'fixed' distant stars. Forgetting the Sun, the sidereal day is one complete spin of the Earth.

Note the difference of nearly four minutes between the two days. Now, as far as I am aware, there is no name for this time difference. Yes, indeed, it's a problem for me. I, and I am unanimous in this, hereby name the four-minute time difference between the solar and sidereal days the **Vamplew Time Shift** (VTS). I've always wanted a time-unit thing, and now I have one.

To really understand this near four-minute shift, or rather the VTS, you could do a simple experiment. Stand up and pretend that you are the Earth. Be as real as you possibly can: wear blue (about 70 per cent to represent the sea), white (for clouds, snow and big sheep), brown (deserts and brown things) and some green

Anna - the Sun
(fixed on the spot)

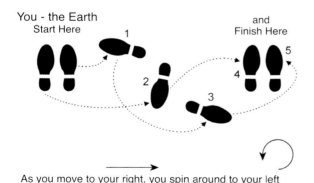

You - the Earth
Start Here

1

2

3

4

5

and
Finish Here

As you move to your right, you spin around to your left exactly once

Let's all do the Sidereal Five-Step to celebrate the spin of the Earth. This small dance illustrates why, after exactly one turn, any place on Earth does not come back to face the Sun again. Anna may tap her feet at any time if she's getting bored.

(trees and grass). Next you need a friend, preferably called Anna, to pretend to be the Sun – make sure she 'shines' appropriately. To reproduce a day, start by facing each other a few metres apart. Anna, the Sun, stays still, while you spin around once anticlockwise (to the left) on the spot, coming back to face the Sun again. You have just done a day, and how easy was that! Try it again if you're enjoying yourself.

Now, here's the thing, that isn't really how it happens. Turn around again, but this time move to the right as you turn. Remember to turn exactly once, that is a whole 360 degrees, but no more. Super, you have successfully represented the Sidereal Day. What should be apparent is that after your spin you are no longer facing Anna – she is off to the left a bit. Well, of course she is, you have stepped to the right. In order to face Anna again you must make an extra small anticlockwise turn. For the Earth, this little turn is that near four minutes' worth of spin. What you did by stepping to the right was to simulate the movement of the Earth in its orbit, ending up with that necessary extra bit of turn (the VTS) to get back to facing the Sun.

Now, hang on a cotton-pickin', rootin'-tootin' minute! Our lives are run by the Sun, clocks and everything – what has the Sidereal Day really got to do with anything?

Well, not a lot with our daily lives, but as far as the constellations are concerned, this four-minute shift has noticeable effects on the night sky.

If I were to watch the constellation of Orion from Bad Goisern in Austria every day at exactly 7pm, I would first catch a glimpse of the entire group in mid-December over in the east. The

15 Dec **7pm**

15 Jan

28 Feb

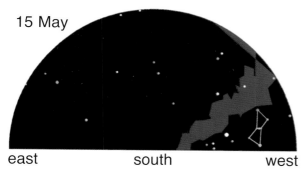

15 May

east   south   west

The positions of Orion, the Hunter, on a variety of dates from mid-December to mid-May, always looking south and always at 7pm. The slow shifting of constellation positions is a consequence of the Earth orbiting the Sun. The blue band flowing across the sky is the Milky Way, although the section near Orion is the faintest and may not be noticed. (Charts created by Scientific Astronomer software, Wolfram Research, Inc., Mathematica, Version 5.2, Champaign, IL [2005])

following day, it would have shifted slightly and would now be seen to be a little higher in the sky and a bit further south than the previous night. Its highest position, due south, would arrive at the end of February. After that, day by day, Orion moves down to the west and starts to disappear in the middle of May. Remember that I am charting Orion's position in the sky only from a fixed point in Austria at 7pm each day. But what applies to Orion applies to every constellation, and to the entire sky, wherever and whenever you are looking at it.

Using the real solar system, but considering an alternative reality, let's look at how this all ties together with the different spins and moving constellations. We would have a much easier time if the Earth didn't move – so forget the physics stuff, let's just imagine… imagine… imagine… Some nasty critter, probably from the Andromeda Galaxy, has stopped the Earth from orbiting the Sun, but left it spinning. At astronomical midday we face the Sun (it's at its highest in the daytime sky) and at astronomical midnight we face directly away from the Sun. Under these circumstances we will always see the same constellation at midnight because we are not moving. Nothing will ever change at night – we will always see the same constellations rise at the same time.

Now add a splash of orbital motion and things begin to change. In one spin we have travelled a small way along our orbit and there's the extra little bit of turn to get us back to facing the Sun (at midday). Not only that, but the direction away from the Sun (at midnight) has slightly changed too. It's this orbital shift that has made Orion, and every other star, change position day by day.

Oh no, some crazy Andromedan UFO has stopped the Earth from moving around the Sun. We're stuck – still spinning, but stuck in place. There's a red dot showing the place where astronomical midday is, and another showing where it's midnight. Notice that Orion is slap-bang in the middle of the midnight sky. As the Earth spins on its axis, Orion will always be there in the same place at midnight. After disabling the Andromedan's ray we can get back on course... (Orbit plotted by Scientific Astronomer software, Wolfram Research, Inc., Mathematica, Version 5.2, Champaign, IL [2005])

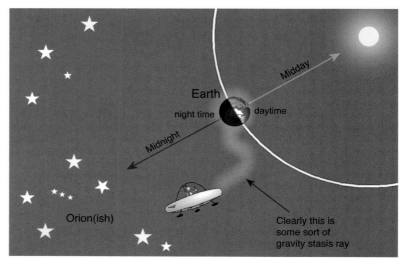

NOTE: THE DIAGRAM BELOW IS HIGHLY EXAGGERATED FOR THE PURPOSES OF CLARITY.

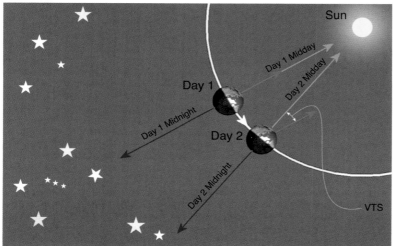

Travelling along our orbit, we can see that the direction to the Sun (midday) and the direction away from the Sun (midnight) change day by day. Day 1 midnight looks directly at Orion, but Day 2 midnight doesn't – it's shifted. Now notice the green arrows. Imagine one green arrow actually sticking out of the Earth like a flag-pole. That's Day 1. As the Earth spins, the arrow, of course, is carried around. When it's pointing in exactly the same direction again we know the Earth has spun exactly once, and Day 2 shows the arrow after one revolution. But wait, our green arrow is no longer pointing at the Sun. That's what the Earth's orbital motion has done. In order to get the arrow lined up with the Sun we need to spin the Earth a bit more – up to the Day 2 midday arrow. This extra bit is the VTS, or 3 minutes 56 seconds, leading to a near four-minute drift of the stars from one night to the next. (Orbit plotted by Scientific Astronomer software, Wolfram Research, Inc., Mathematica, Version 5.2, Champaign, IL [2005])

If you continue this idea around the Earth's entire orbit, then this small daily shift will, in the course of a year, enable us to see all the stars that are at some stage hidden in the daytime sky. It's like standing in front of the Leaning Tower of Pisa – you cannot see what's on the other side of it (obviously, because it's in the way!). Similarly we cannot see the stars on the other side of the Sun, because for one thing the Sun is in the way and for another our atmosphere is bright blue (or cloudy). As you walk around the Leaning Tower, though, you begin to see what is on the other side. So as the Earth moves around the Sun, the stars that have been visible in the daytime slowly disappear into the night.

So, answering our initial question, this is why we see different constellations at different times of year.

Well, I got there in the end, but you see what that question did? The answer went off on its own cosmic course of days and orbits; I even managed to get a new time unit in there.

CHAPTER 2

A Year in the Life of the Earth

The ancient Earth-centred version of the Universe as depicted by some of the great Greeks, including the astronomer Claudius Ptolemy and the philosopher Aristotle.

In the previous chapter, we looked at the daily spinning Earth and our orbital motion around the Sun. All of which leads very nicely into the realms of earthly tilts, solstices, equinoxes, inclinations, the zodiac and the ecliptic – great astronomical words for things that affect how we see the night sky through the year.

Let's start here with a sweeping statement: the Earth goes round the Sun. Our speed in orbit of 30km per second is a consequence of the distance our planet was from the Sun when it was made, balanced by the Sun's gravitational pull. With the greatest effort in the world there is no way to feel this motion, just as we cannot feel the Earth spinning on its axis. We know, because we have been told, that the Earth goes around the Sun. But what if really the Universe revolved around us?

Up until the early 1600s, the idea that the Sun went around us was accepted without question by a lot of people, and it took the invention of the telescope and advancements in mathematics to show that this might not be what was actually happening.

The reason for believing this geocentric (Earth-centred) model was simply that it mirrored what we saw happening in the sky. Indeed, I don't think it would make much difference to our lives if we continued to believe it today. Taking an Earth-centred view, it appears that the Sun moves around us over the course of a year. It's a nice smooth movement that at times speeds up or slows down a little depending on the slight variation in distance between the Sun and us. If we think of the Sun as an athlete running around a track, while we, the Earth, stand watching in the centre of the stadium, and if that athlete sprays some blue-coloured paint in the air that just 'hangs about a bit' while running, we will see a nice circular line all the way round us.

In the real solar system this invisible (blue) line is called the **ecliptic**, a word that means 'place of eclipses,' for when the Moon and Sun are lined up correctly on this line, we see...you guessed it, an eclipse.

It's the constellations that lie behind this ecliptic line that have become woven into folklore. This is a twelve-strong group of figures collectively known as the **zodiac**. By definition the Sun spends some time moving across each zodiacal constellation – there's one per month, although the actual time you'll find the Sun in each varies considerably, as the table below shows.

Virgo wins hands down, with the Sun spending a whopping 44 days travelling across her. But look at Scorpius with the Sun here for only seven days. The italicized group Ophiuchus hardly gets a mention, when it should really be the thirteenth member of the zodiac simply because the ecliptic (the Sun's path) lies across it and also because the Sun spends more time in it than in Scorpius. Hocus-pocusly, if your birthday is between 30 November and 18 December, then you should technically be an Ophiuchean. Now, how marvellous is that?

Take a look at the Starry Skies map below to see the ecliptic and official zodiac in its full glory.

Forgetting the grid lines and additional information for a moment, you may be wondering after all this

Zodiac constellation (except the italicized one)	Date the Sun enters the constellation	Number of days the Sun spends in the constellation
Pisces, the Fish	12 March	38
Aries, the Ram	19 April	26
Taurus, the Bull	15 May	37
Gemini, the Twins	21 June	30
Cancer, the Crab	21 July	21
Leo, the Lion	11 August	37
Virgo, the Maiden	17 September	44
Libra, the Scales	31 October	23
Scorpius, the Scorpion	23 November	7
Ophiuchus, the Serpent Bearer	*30 November*	*18*
Sagittarius, the Archer	18 December	32
Capricornus, the Sea Goat	20 January	26
Aquarius, the Water Carrier	16 February	24

Based on the years 2010 – 2011

This Starry Skies map has been warped to work on a flat page, when really it's a three-dimensional sphere all around us. There's a lot going on here, so I'll come back to it on several occasions in the course of the book. The blue line is the ecliptic – the path of the Sun through the year. You can see how it travels through the twelve famous constellations and in so doing has formed our zodiac. (Ecliptic and stars plotted by Scientific Astronomer software, Wolfram Research, Inc. Mathematica)

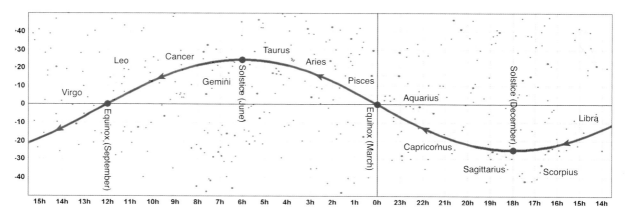

circle-round-us business why the ecliptic is a wavy line on the map. Well, we get this effect because of the necessity of making the whole sky flat, together with the co-ordinate system (see page 13). But if you look at the cut-out and curved Starry Skies map, right, you can see that it is a circle and that the ecliptic is truly round. Imagine the Earth being in the centre of this Sun-path circle (just like us with the athlete on the track) and it will start coming together quite nicely.

Of course I've tilted the cylinder so you can get the best view of the ecliptic inside and out, but fascinatingly it's leaning over at 23.5 degrees – fascinating indeed, for that number represents the tilt of the Earth's axis. Taking a side-on view of this cylinder (yet another great diagram) shows the tilt more clearly and how it is identical to that of the Earth's axis.

Here's the cut-out and curved Starry Skies map. To save you taking your scissors to the book to cut out the map, curve it round and get this, I've done it for you. The ecliptic (thick blue line) can be seen to form a circle around the sky.

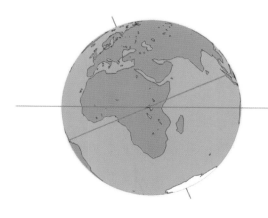

Ecliptic

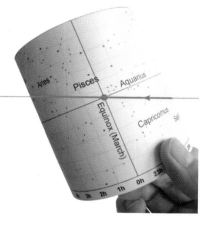

A side view of the leaning cylinder next to the Earth. Here the ecliptic appears as the line right across the image, with both the Earth and the cylinder tilted over towards it at 23.5 degrees. NOTE: it is the Earth that is responsible for the tilt of the cylinder; if we changed the angle of the Earth's axis then the cylinder would change correspondingly.

So now we know that the reason those marvellous old Earth globes that sit in museums are tilted is that that is how the Earth is 'fixed' in space in relation to the ecliptic – and therefore the Sun. In advanced astro-speak this is known as Earth's *obliquity of the ecliptic* – a good set of words to use in those intelligent moments.

This 'fixed' tilt means that the same stars appear above the poles of the Earth all year round. We can see this in the northern hemisphere by the fact that Polaris (the Pole Star) is always in the same place at night – as it's virtually directly above the North Pole. If the pole was shifting anywhere over the year as the

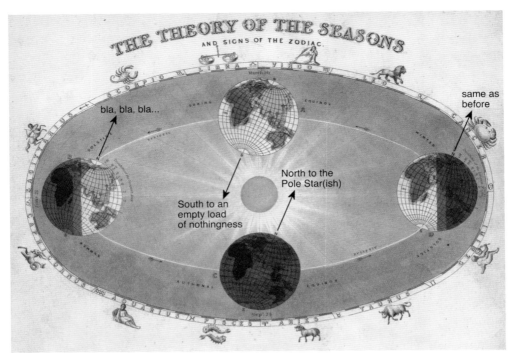

THE THEORY OF THE SEASONS

AND SIGNS OF THE ZODIAC.

bla, bla, bla...

South to an
empty load
of nothingness

North to the
Pole Star(ish)

same as
before

This is why we have seasons – the Earth's changing tilt in relation to the Sun. However, there is no shift in relation to the distant stars – no matter where the Earth is during its yearly orbit, the North Pole always points very close to Polaris, the Pole Star. As you can see, there is no equivalent southern 'Polaris Australis' star above the South Pole – it's all just a little too empty, dark and scary for my liking.

Earth travelled round the Sun, then we would see a change in Polaris's position. (Technically there is an interesting wobble called *precession* that does slowly move the pole, but seeing as it takes 25,800 years to complete, it is not going to affect how we see the night sky considerably during our life time – so let's forget I even mentioned it.) What was I talking about?

Oh yes. With the tilted Earth going around the Sun, over the course of a year we can start to make sense of the astronomical points that mark the start of the seasons. Starry diagrams showing this invariably look something like the old one shown below, with four Earths at various positions around the Sun.

A lot of people believe the seasons are caused by the distance the Earth is from the Sun. It is true that the Earth does have a slightly elliptical orbit, leading to a difference of 5 million

kilometres between its closest point to the Sun (known as *perihelion*) and the furthest point (*aphelion*). Sounds a lot, but it is not this that affects the seasons. In fact, during the northern-hemisphere winter, the Earth is as close to the Sun as it can get (perihelion occurs around 4 January).

So, we're back to this tilt and the four-worlds diagram above to find the explanation. Taking the Earth on the left of the diagram first, you'll see the North Pole tilting as much as it can (23.5 degrees) towards the Sun, while the South Pole is tilting as much as it can away from the Sun. The moment in time when this occurs is called the **solstice**. For the northern hemisphere the day this happens marks the first day of summer, and for the southern hemisphere it's the first day of winter (around 20 June). There's another solstice point six months later (the Earth on the right), when the tilt of the poles is completely reversed in relation to the Sun: now the South Pole is sunward and the North Pole leans away into space. This marks the start of the northern winter and southern summer (around 21 December).

In between the two solstices are Earths where both poles have no tilt relative to the Sun. At these moments we have the **equinoxes**, which mark either the start of the northern spring and southern autumn (around 20 March), or vice versa (around 22 September).

The positions of the solstices and equinoxes are marked on the Starry Skies map on page 10, and now is the correct time to bring the grid into play. Just as latitude and longitude are used to find places on Earth – for example, Lat: 47 degrees 38 minutes North, Long: 13 degrees 37 minutes East describes the location of Bad Goisern in Austria on the Earth – the same idea works with space, except that here latitude is called **declination** (or **dec.** for short) and longitude is called **right ascension (RA)**.

Declination shouldn't give us any bother as it uses the usual + degrees and minutes (with the numbers getting bigger as you move into the northern hemisphere 'up' from the equator) and – degrees and minutes (for movements into the southern hemisphere), just like on Earth.

However, we need an extra step for right ascension, as it uses hours and minutes. One hour of RA is the equivalent of 15 degrees of longitude – that's the left or right movement around a map. This is because in one hour the sky turns 15 degrees. So a day's movement is 24 hours' worth of turn x 15 degrees, which equals 360 degrees – all the way round once.

As an example, the solstice in June on the Starry Skies map can be located with the co-ordinates 6h 0m RA, +23° 30′ dec.

The zero-degree declination line that runs across the centre of the map (see page 10) is known as the *celestial equator*, and this represents a projection into space of the Earth's equator. Easy declination, once again.

RA, of course, decides to be awkward: the vertical zero-hour right ascension line is marked at the (equinox) moment the Sun crosses the celestial equator. The forgotten precession from earlier means that this point and its associated line, and therefore *all* lines on the map, shift very slightly each year. So when positions are given for astronomical objects in

Right Ascension. 1900.	Declination. 1900.	
h. m.	° ′	
5 30	− 5 28	Struve found another very faint star (10 mag.), d 184°; and another of 10·5 mag., distance 149″, po Σ 747, **Orionis**. A double star, 5·6 and 6·5 position of the companion has not changed since 1179 (1976), **The Great Nebula in Orion.** interesting object of its kind in the Northern hea field of research to all who have a powerful t No description can give an adequate picture of accounts given by De Vico Herschel and others

ORION NEBULA   NGC 1976 + M42

R.A. (2050)
5h 38m
dec.
−5° 21′

The Orion Nebula is the famous faint cloudy patch you can see with your unaided eye sitting below the three belt

The top image is a section of a page from an old star atlas that gives the co-ordinates of the Great Nebula in Orion for the year 1900. Below that, a newer publication shows its position 150 years later. The nebula hasn't actually moved, of course, only our map reference grid.

Eris 44°
Pluto 17.2°
Mercury 7°
Venus 3.4°
Saturn 2.5°
Mars 1.9°
Neptune 1.8°
Jupiter 1.3°
Uranus 0.8°
Earth 0°

Sun

Planetary inclinations in relation to the ecliptic. Of course the Earth's orbit has a zero-degree incline, as the Sun–Earth line ecliptic is our solar system base. The higher the inclination, the further the planet can appear from the ecliptic.

space, they are only correct for a particular date – although as mentioned, the shift is extremely slow. However, if you do have any old books on astronomy, you may not be able to find faint objects, as you'll be looking in the wrong place.

We also use the ecliptic as a base line, or reference 'level', for the solar system. Back to the athlete, but this time we're more interested in the track itself – it's nice and flat around us. If we add a few more runners and set them off at different speeds it won't be long before they're spread out round the track. But simply because they are *on* the track, they are all on a circular line at the same level around us – you can say that everyone is on the same *plane*. This is how we see the Sun, planets and Moon moving in the sky: all appearing on the same track – the ecliptic. Of course, some planets are closer or further, something impossible to detect with the eye, but they still appear on (or near) this line.

Actually, unlike the racetrack, which is on the same level thanks to the ground, of course there is no ground in space, and the 'track' comes from

the fact that the solar system was made out of a wide disc of material. Interactions and collisions of bodies within the disc formed the planets and caused some variation in how orbits are inclined to the ecliptic, so although the planets basically follow this path, some can be found above it or below it, depending on their place in orbit.

The diagram above shows how other planets' orbits are tilted in relation to the ecliptic – that is, their *inclination.*

And, on a final point about the grid, the ecliptic and the positioning of objects in RA and dec.: the cylinder, of course, should really be a sphere all around us. We call this the *celestial sphere*, with the Earth nestling safely in the middle. Due to technical difficulties, producing a map you could cut out and make into a sphere was not a responsible thing to do, hence the cylinder. But if you go out at night, it's not hard to imagine this sphere around us onto which everything is projected – no matter how near or far objects are, they all appear to be on this sphere.

CHAPTER 3

The Sun

Our star, the Sun: what an incredible story of creation and existence it is. If we look at how the Sun was born, then it's possible to widen the story to include all the stars we see in the night sky, because they're all made the same way. To begin our tale, we must go back to around 6 thousand million years ago when the Sun hadn't even been thought of. Our area of the galaxy was filled with a vast cloud of dust and gas – a nebula – drifting through space. Maybe a nearby star exploded, sending shockwaves through the nebula one day, or perhaps it encountered another cloud. In either case, the result was compression of dust and gas within our nebula. Now, once you've pushed enough stuff close enough together there's no stopping a certain overriding energy – parts of the nebula *will* start clumping together under the force of gravity.

As things pull together, the heat begins to build and if you have the right amount of gas – not too little, not too much – then you're on the way to creating a star.

Clumps such as these that we see forming in nebulae today are known as *protostars*. If there is enough material, the temperature is able to reach 10 million degrees Celsius. This is the magic figure because at this temperature the most important nuclear processes can start in the gas in the centre of the clumps – this is when a star is born. For the Sun to get to this point from the wispy shocked nebula stage above took just under 500,000 years.

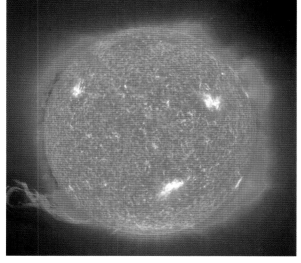

Our very active, massive Sun – 1.3 million times bigger than the Earth.

(Image courtesy ESA/NASA)

1. Nebula

2. Gravitational Clumping

3. Proto-Planetary Disk (Proplyd)

4. Solar System

Our solar system started life in a nebula (1). Clumps formed due to gravity (2), one of which became the Sun with its disc of dust and gas (3) that slowly accreted to form the planets and rubble (asteroids, comets, etc.). (Image courtesy π Studios)

Of all the stuff in a star-forming nebula, the predominant ingredient is hydrogen gas. Basically hydrogen is what the stars are made of and what they use for fuel – in other words, they are their own fuel. In the nuclear furnaces of stellar cores, hydrogen is converted into another gas, helium, releasing a lot of different kinds of energy in the process. Among the most important is the energy of pressure. It's this force that stops the continuing gravitational collapse of the gas that was happening in the *protostar* stage, and then keeps the star 'up'. This is very much a balancing process that took around 30 million years for the Sun to complete. Once shining away happily the Sun became a standard *main-sequence* star, a name given to all well-behaved stars and the one by which they are known for a large percentage of their life. And so

NEBULA PLURAL **nebulae** | **Latin for *mist* or *little cloud***

The clouds of potential star-fluff we call nebulae can take on different guises depending on how much dust and gas there is, how they're made and what's nearby. Given enough time, any of the following can merge or split to become one or more of the others – there are no hard and fast rules here.

The **Orion Nebula**. (Image courtesy Paul Whitmarsh)

EMISSION NEBULA

Associated very much with the birth of stars, an emission nebula is a cloud that is 'glowing' from radiation supplied, for the main part, by the stars made within the cloud itself. A lot of pictures of these things show mainly red wisps and swirls, red being the colour emitted by hydrogen – the main gas of a nebula.

The **NGC 1333** nebula in Perseus. (Image courtesy NASA/ JPL-Caltech/ R. A. Gutermuth)

REFLECTION NEBULA

Sometimes a cloud is just hanging about not doing a lot (yet), but it is dense enough to reflect the light from any nearby star. A reflection nebula tends to be fainter than an emission nebula, and as some of the light is scattered by the cloud, just like the sunlight through our atmosphere, it can take on a bluish colour. Of course you have to take into account the colour of the nearby star, too, as it will reflect off the nebula, thus changing its coloured appearance – just as the golden red of the setting sun reflects off a white wall making it appear golden.

here sits the Sun today, midway through its life, generating energy from its hydrogen fuel that it is using up at a rate of 4 million tonnes per second.

Now deep in the Sun, because there is so much compressed and excited gas, it can take the energy millions of years to bounce and weave its way up to the 'surface', before it is radiated away into space at the speed of light. The energy at the time it has left the Sun includes what we see as sunlight and feel as heat, but don't think it started that way. Let's travel to the core to see what's going on.

The *core*: this is where all the action is. I mentioned 10 million degrees Celsius, but that is just the minimum you need to get a star going – the Sun is working with a substantially warmer 15 million degrees. It's also very hard to move down here, as the density is ten times

DARK NEBULA

How do you see something that's dark? By putting it in front of something that's bigger and brighter. This is how a dense-enough cold nebula gives itself away. Take the classic Horsehead Nebula (B33) – a wonderfully shaped dark cloud silhouetted against the bright nebula (IC 434) behind. Other dark nebulae can be seen along the Milky Way, as they block out the light of the stars behind.

B33, the famous **Horsehead Nebula** in Orion. (Image courtesy Equuleus Studios)

PLANETARY NEBULA

This is a process of a star returning its gas (a lot of it changed from how it started) back to the Universe for recycling. As stars like the Sun approach the end of their lives they swell and puff off their atmospheres, which if you are looking at them through a telescope look similar to planets – hence the name. Magnetic fields, companion stars and other factors can make planetary nebulae appear incredibly intricate when studied closely.

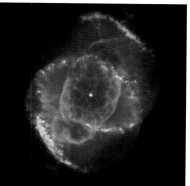

The **Cat's Eye** planetary nebula in Draco. (Image courtesy NASA/ AURA/STScI)

SUPERNOVA REMNANT

While a planetary nebula is a 'gentle' push-away of an ordinary(ish) starry atmosphere, a supernova is the final all-out explosive destruction for a massive star. There's nothing nice and round floating away here; it is all messy, rough and spidery-looking gases crashing outwards into the Universe at tremendous speeds.

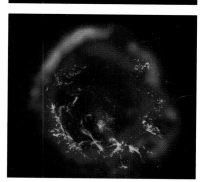

Supernova remnant E0102 in the Small Magellanic Cloud. (Image courtesy NASA/CXC/ SAO/HST)

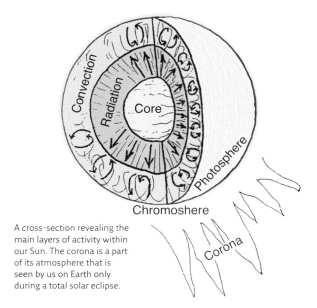

A cross-section revealing the main layers of activity within our Sun. The corona is a part of its atmosphere that is seen by us on Earth only during a total solar eclipse.

that of lead. Extreme all round, then, and that goes for the energy too: being released in the core in the form of ghostly particles called neutrinos, gastastic gamma radiation and xylophonic X-rays. Without a shadow of a doubt, these energies are not good for the health, so thank goodness that by the time they've travelled through the various Sun layers they've calmed down to the nice lighty-stuff that makes our planet so sunny.

A journey out from the core involves firstly a visit to the surrounding *radiation zone*, the job of which cannot be underestimated – that is to keep the core going. Thanks to the radiation zone maintaining the pushing all-round gravitational pressure, the core temperature will stay up there where it needs to be.

The energy in this radiation zone has found itself in a lower, 4 million degrees world – a number that continues to fall as the distance from the core increases. This cooler temperature makes the energy somewhat *sticky* – that's because it's easy for energy to move by radiating

itself through very hot gas, but this slightly less hot variety is not so friendly. As we all know, there's nothing worse than sticky energy and the result is it heats up the surrounding gas so that it (the gas) starts moving and the energy can use it (the gas) to transport itself (the energy) out.

We've arrived at the *convection zone*. START: it's a place where currents of hot gas, known as cells, are constantly rising as they are heated from below. Each cell splits up a few times before it reaches the bright *photosphere* (the visual surface of the Sun), where the top of the rising column is seen as a granule. The heat, light and energy have made their final journey and are finally released into space. The now cooler cell gas falls back down to be warmed again. Go back to START and repeat constantly for another 4.5 thousand million years.

The temperature in the photosphere is just under a cool 5,500°C, but with all the hot stuff churning away underneath it's not surprising that there's plenty to look at – with safety. So, as we venture into the realms of observing features 'on' the Sun, yes, here come the usual warnings:

DANGER! DANGER!
HIGH BRIGHTNESS AND HEAT!
STARS ARE NOT A TOY AND WHEN YOU'VE GOT ONE CLOSE BY (LIKE THE SUN) EYE DAMAGE WILL RESULT UNLESS YOU'RE VIEWING FROM A SAFE DISTANCE: AT LEAST 6 LIGHT-HOURS IS THE RECOMMENDED MINIMUM.

THE EARTH IS ONLY 8.3 LIGHT-MINUTES AWAY FROM THE SUN, MAKING IT A BIG VISUAL NO-NO WITHOUT PROPER PRECAUTIONS.

From all of those creation scribblings we can now better understand the features we see when we safely observe our Sun.

Photosphere Features

Don't think of the photosphere ('sphere of light') as a surface as such, for it has a certain depth to it, much like looking into the water and seeing all the way to the bottom of a clear swimming pool. Within this photospheric 'pool' of semi-transparent gas the attractions to be found are:

Granules

Virtually the entire photosphere, except for the sunspots that we shall consider below, is covered in the granules that featured above. They range in size from around 500 to 1,500km across and last maybe 20 minutes before new hot columns push the old ones out of the way at speeds of up to 24,000kph.

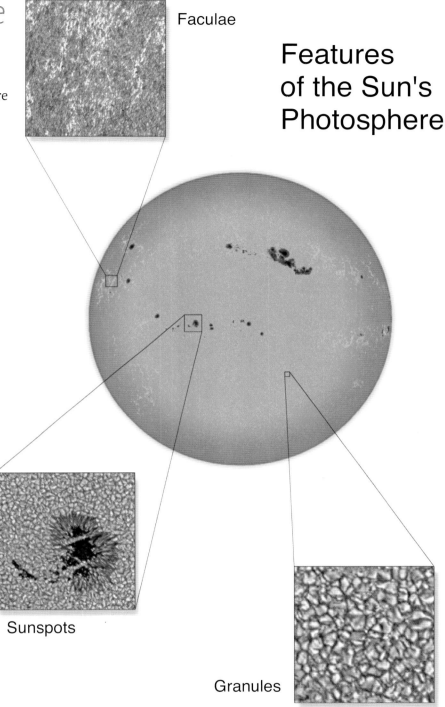

Faculae

Features of the Sun's Photosphere

Sunspots

Granules

Sunspots

At times, lower temperature patches called sunspots appear on the bright photosphere, lasting anything from a few days to a few weeks. How many there are at any one time depends on where the Sun is in its solar cycle – this is an eleven-year-ish magnetic wind-up caused by the Sun spinning at different speeds. You see, the Sun's equator rotates around once in 25 days, but go to the poles and that's slowed down to 34 days. You can blame these changing rates of the Sun's 'day' on the fact that it is not solid but gas.

As for the 'magnetic wind-up', imagine that moving invisibly with the Sun is a network of secret magnetic fields: lines that initially run directly from the North to the South Pole. They're buried with the rotating (at different speeds) gas, but like rubber bands they can stretch. As the equator moves ahead, so the magnetic fields are pulled out longer until they can't take it any more and they snap, warp or join up with another nearby field – this is near *solar maximum* at the end of the eleven-year cycle.

One of the results of these disturbances could be a sunspot. All this magnetic disorder stops energy reaching a patch of the photosphere, which makes it cooler to the tune of 2,000°C.

Sunspots can appear on their own, but more commonly in pairs or in two groups of spots. As it's all magnetic effects going on, just like a bar magnet, one of the groups will be positive (that is it will have a positive field) while the other is negative.

After solar maximum the fields have all realigned themselves north–south, and the process starts again. There is one small change: after every cycle, the North and South Pole of the Sun swap positions.

Faculae

These are small bright features on the Sun, associated with sunspots, created by magnetic forces in the gaps between granules. They may be small, but their vast numbers, much more than sunspots, can actually make the Sun slightly brighter, especially when near solar maximum.

The equator of the Sun rotates faster than the polar regions by about nine days – arrow lengths indicate these different rates of spin. The red line is one of the many magnetic field lines that starts its life running straight from the North to the South Pole but which gets stretched by the differential spin, wrapping its way tighter and tighter around the Sun. Such winding cannot last forever and slowly, over a period of about eleven years, the Sun becomes very active as the wound-up fields begin to connect together and smash through the surface of the Sun. This is the solar cycle, with the most active period known affectionately as 'solar max'.

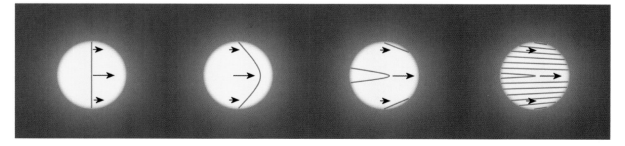

An area where faculae develop could see a sunspot within a few days. Later, after the spot has gone, faculae sometimes hang around for months. They tend to be more noticeable towards the edges of the Sun, where the photosphere is less bright.

With two bits of handy-sized cards I made earlier I can see sunspots and eclipses in super-safety.

Observing the Sun

With nothing more than two pieces of card, one with a hole in, you can see large sunspots – if there are any! Let the sunlight fall through the hole in one card onto the second. By varying the distance between the cards you can focus the light until you see a crisp outline, and hence the best views of the spots.

With a good telescope (except the mirror-type reflectors, the eyepieces of which get very hot) or binoculars, the safest way to observe features of the photosphere is to project the Sun's image onto a large piece of white card. However, do not attempt this until you are confident that what you are doing is safe. If you are using a telescope, make sure that the finder is suitably capped, and the same with one of the lenses if you are using binoculars. Otherwise focused sunlight will cause some nasty effects.

You can line up the telescope with the Sun approximately by watching the telescope's shadow on the ground. When the shadow is at its smallest, with any luck the Sun's image will be coming down the scope – you may need to 'wobble' the scope a little to get the perfect position. This all applies to binoculars as well.

Instead of your eyes being at the eyepiece, you place a piece of white card there. The further the card is from the eyepiece the larger the image, but also the dimmer it gets – so this is a balancing act. When you've got it right, after focusing, you and all your excited friends can see and draw the sunspots.

Some cheaper telescopes are supplied with sun filters that fit onto the eyepiece and supposedly let you look directly at the Sun through the telescope. Throw these away immediately. They are very dangerous because they take the full focused power of the Sun and could shatter. *Do not use eyepiece sun filters.*

Experienced amateurs buy 'mylar' filters that fit over the big end of their scopes. These cut out all the infra-red (heat) and ultra-violet radiation from the Sun and 99.9 per cent of the brightness, so they do allow a safe direct view through the telescope, but again, use them only if you know what you are doing.

CHAPTER 4

Stars and their Life in General

The basic difference between the Sun and the stars we see at night is merely one of distance. Those twinkly night-time things are so much further away than the Sun that they look like tiny dots of light. Some bright stars may seem to be bigger, but that is only down to your eyes. For some of us it's also our eyes that can make stars look like the classic five-pointy shape, but there's no way Mrs Gravity is ever going to let that exist in space. (I've asked her, and she is really adamant about it.)

What you can notice with a particularly bright star is – if it has one – a hint of colour. Some fine examples include the goldish Aldebaran, the yellowish Capella and the bluish Achernar. There are actually loads of coloured stars out there, but they are just too faint to get the colour-receiving cones in our eyes going – that's why most look white. Actually, a lot of this comes down to eye dark-adaption, light pollution and how long you spend gazing in wonder at the night sky. The gentle hues of star colours sometimes need time to reveal themselves, so next time you're out on a particularly clear night, just pick a few bright

A multitude of different stars fills the cosmos. How big and what colour a star is depends on how much gas made it in the first place and where the star is in its life. The Sun is a pretty average size, but there are many larger varieties – although not all of them are at the same stage of life, so it isn't fair to compare them directly.

stars, compare them and see if you can see any even slight coloury tinge.

If we leave our Earthly viewing on a home-made rocket and get up close to another star, then we can see and measure any other differences between it and the Sun. Those differences will be in size and the amount of energy being produced. This is all because of one thing (I mentioned it in the last chapter, but now for the details): how much gas went in to making the star in the first place. To put it simply, more gas makes a bigger star – that is, one with more mass.

Bigger, in starry terms, also means brighter – the temperature in the core of a star larger than the Sun is higher and so more energy is made all round. It's just like a car with a big engine as opposed to a small one. The big engine produces more power. However, it also uses up fuel at a greater speed – and stars with more mass than the Sun do exactly the same thing. Now, if you're going to go around using up your fuel quickly like a large star, you're not going to live as long – live fast, die young. Whereas the Sun's life is estimated to be around 10,000 million years, massive Rigel only has 100 million years. Conversely stars smaller than the Sun can last for a trillion years or longer.

Star Colour	Colour Class (spectral type)	Temperature (°C)	Mass	Absolute Magnitude	Real Starry Example
	M	1,700 to 3,500	0.1	7.5 to 15	Proxima Centauri
	K	3,500 to 4,800	0.8	5.5 to 7	Epsilon Eridani
	G	4,800 to 5,750	1	4.25 to 5.25	Sun
	F	5,750 to 7,000	1.7	2.75 to 4	Polaris (UMi)
	A	7,000 to 10,000	3.2	0.5 to 2.5	Sirius (CMa)
	B	10,000 to 30,000	18	-5 to 0	Regulus (Leo)
	O	30,000 to 60,000	60	-11 to -5	Alnitak (Ori)

So, what exactly do you get for your money? Here is a guide to what star you'll end up with depending on how much gas (*mass*) you've got to make it in the first place. The figure in the mass column represents how many times more or less gas the star has compared to the Sun, and this determines what 'surface' temperature and colour your star will be.

The basic fact we can get straight from the table is that temperatures and colours range from hot blue stars down to cool red ones.

Each of the colours has an associated letter. Stars are often mentioned by their class (*spectral type*), so if you hear an astronomer say, 'It's a B2 type star', you'll know that's just a posh way of saying, 'It's blue.' Yes, indeed, there's a number in there too, as each class is divided into ten, e.g. M0, M1...M9. This is because there isn't really a single colour for each class, but they change slowly from one to the next one as the mass (and hence the temperature) changes.

Using the details from the table, M0 would be the hottest star of its class with a surface temperature of about 3,500°C and an orangey-red colour. Here we've just hopped over the rope from the end of the orange K9 stars. By the time we get to M9, we find the coolest stars at around 1,700°C with a fine red colour.

On the mass side of things, there is a lower and upper limit to the sizes of stars. Below 10 per cent (written as 0.1 in the table) of the Sun's mass (also referred to as *solar mass*) there starts to be not enough gas pressure to raise the temperature high enough to get that hydrogen nuclear engine going. At a very low 8 per cent (0.08) solar mass the object made is called a **brown dwarf**, or sometimes a 'failed star', as only lukewarm non-hydrogen reactions can occur. Here is the murky boundary between stars and large planets. It's just a matter of size.

The Christmas Tree Cluster (NGC 2264) in Monoceros is where it's all happening – stars of all sizes are being made from the surrounding gaseous material. The marvellously shaped cone nebula sits right at the bottom of the scene. (Image courtesy T.A. Rector/B.A. Wolpa/ NRAO/NOAO/AUTA/ AUI/NSF)

Variable Stars are stars that change in brightness (and possibly colour (a bit))

There are two types:

1. INTRINSIC VARIABLES

Stars at the start of their lives, before they've had a chance to settle down, or near the end of their lives, after they've had enough of the settled-down existence, do funny things. One of those things could be a pulsation in size with a corresponding change in brightness and colour. So, intrinsic variables are stars that actually change in brightness.

2. EXTRINSIC VARIABLES

This is where our line-of-sight affects a star's brightness in the sky. For example, when two stars orbit each other very closely – and from our viewpoint one 'eclipses' the other – a slight change in the overall brightness of the system is seen. Close systems may also cause a star to be gravitationally pulled into an egg shape – here the changing aspect of the shape as it rotates affects the brightness (side-on egg versus end-on egg) – or indeed cause one star to suck gas from another, with disastrous consequences (see picture below).

One type of extrinsic variable star is the Nova. Here the strong gravity from a white dwarf (the collapsed remains of a old star) pulls off gas from its large binary companion. Once enough gas has built up there is a nuclear explosion, which flings matter into space and causes a brightening of the star, typically by ten magnitudes.

Depending on the type of variable, the period of brightness change can be regular or not, varying from hours to years.

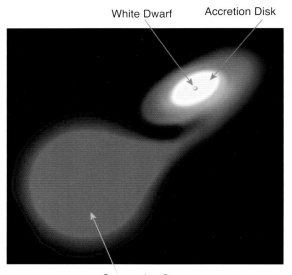

White Dwarf   Accretion Disk

Companion Star

Looking larger, maybe around 150 solar masses, we get near to the upper limit for a star. With so much gas, the heat produced is phenomenal and the star simply makes too much energy and rips itself apart.

Finally in the table is something mysterious that we have never come across before: **absolute magnitude**. We use magnitude as a measure of brightness for pretty much anything in space – the stars, Sun, planets, comets, the Moon: most things really – but there are different types of magnitude.

How bright something *looks* in the sky is given by its **visual** or **apparent magnitude**. While participating in general outdoor stargazing (as opposed to the indoor variety) you'll hear things like, 'Polaris is about mag. 2.' Deeper discussions about numbering follow shortly, but 'mag.' is an abbreviation for *apparent magnitude*.

The problem with stars is that they are not the same absolute magnitude. Some are thousands of times brighter than others. We cannot see any of this when we look into the sky: a small star may appear bright because it's close to the Earth, while a real super-shiny one may appear very faint only because it is so far away. If only we could somehow line them all up at one distance, then we could judge them equally. But how is that possible?

There is a magnitude equation thing that does exactly that – on paper anyway. What you do is measure the apparent magnitude. At the simplest level our eyes can judge brightness and how one star varies from another quite well. Indeed this is how it was done before special space light-meter equipment was developed. The next thing to do is measure the star's distance. Obviously it's not as easy as using a tape measure. Can't see that working up in space for a variety of reasons.

The first star to have its distance measured was 61 Cygnus by German astronomer and mathematician Friedrich Wilhelm Bessel in 1838. The method he used was called **parallax**. It works by noticing a nearby star's movement against a background of distant 'fixed' stars.

To demonstrate, choose one of your thumbs; it doesn't matter which one, but choose decisively. Your thumb represents the star you are trying to find the distance to. Hold it out upright at arm's length and view it against the background with one eye and then the other – there is a shift. This angle is the parallax. With only my right eye open my thumb covers over Mike's skis propped up against the shed and with only my left eye open I can see them. I tell him they should be in a ski-bag, but he's already onto his third beer and doesn't seem to care.

Throwing the parallax angle and distance between my eyes at a calculator I can work out the distance to my thumb. It's a fairly simple triangulation thing.

This parallax principle works in exactly the same way with nearby stars, except, of course, that our eyes are too close together to notice any stellar shift. Astronomers, and I'm still amazed by this, position their eyes 300 million kilometres apart! This is the distance from one side of the Earth's orbit to the other – two points six months apart (see diagram overleaf). Looking from each of these two places is like looking with one eye and then the other, except of course that for any particular star you have to wait half a year from one open 'eye' to the next. Telescopes on the Earth, or in Earth orbit, are used for this operation, as they are the only instruments capable of detecting the extremely small movement of a star that reveals its distance.

By knowing how far away a star is and its apparent magnitude we can find its absolute magnitude. Basically absolute magnitude is as though we've lined all the stars up at a nice safe fixed viewing distance of 32.6 light-years (don't ask!), and therefore we can directly compare one star's actual brightness with another's. There's a table (overleaf) of apparent magnitude versus absolute magnitude of the stars listed in the table on page 23, with the addition of Vega – an important star, as described on page 27.

Just one point here is those minus numbers (-26.7 etc.). You get used to it, but it can seem a little bizarre that a star of magnitude -3 is brighter than one of -1. Even stranger is a star of magnitude zero being brighter than another of magnitude 6, for example, or indeed that a star of magnitude zero is visible at all. This all harks back to ancient times before telescopes

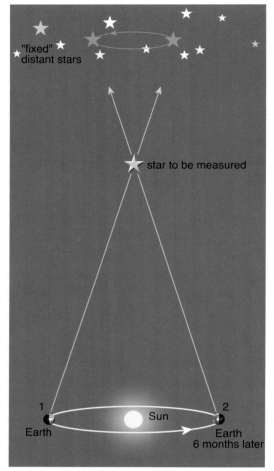

"fixed" distant stars

star to be measured

1 Earth

Sun

2 Earth 6 months later

Here is annual parallax – the change in a star's apparent position over the course of a year. As the Earth orbits the Sun, if we look at a close enough star, it will show a perspective shift against the more distant background 'fixed' stars. The larger the shift, the nearer the star is. From the diagram, the greatest change in the star's apparent position is seen six months after the first measurement when the Earth is on the other side of its orbit.

NOTE: THESE DIAGRAMS ARE VERY EXAGGERATED TO GIVE THE IDEA.

The top image shows how we would see the star in the sky when the Earth in the annual parallax diagram is at position 1 (say, for example, April) and the lower image shows how it has shifted by the time the Earth has reached position 2 (October).

Star	Distance (light-years)	Absolute Magnitude	Apparent Magnitude
Sun	8 light-minutes	4.8	-26.7
Proxima Centauri	4.2	15.5	11.0
Sirius	8.6	1.5	-1.4
Epsilon Eridani	10.5	6.2	3.7
Vega	25.2	0.6	0.0
Regulus	77.5	-0.5	1.3
Polaris	431.0	-3.6	1.9
Alnitak	817.0	-5.2	1.7

when it was the eye that judged star brightness. The Greek Hipparchus started the scale at 0 for the brightest and worked down to 6 for the faintest stars visible.

Of course as technology grew and science became the way of understanding the Universe, much more accuracy was needed. If you're going to grade stars professionally, as with anything, you need a starting point. Where would that have been? I pondered the question for a few moments before I phoned the only person I knew who would have the answer:

Alan. He told me that British astronomer Norman Pogson chose the star Vega in 1856 as the reference star for the new improved apparent magnitude system – it became *Star Zero*.

With the new scale the only thing that could happen with stars brighter than zero was that they slipped into the mystical realm of minus numbers. The table states that our brightest night-time star, Sirius, has an apparent magnitude of -1.4, while, not surprisingly, our brightest daytime star, the Sun, is a whopping -26.7.

As scribbled on page 25, absolute magnitude means a view of the star as if it were 32.6 light-years away. Therefore stars that are closer to us than this distance look brighter (have a higher apparent magnitude) than their absolute magnitude, and conversely stars further away than 32.6 light-years appear fainter (lower apparent magnitude) in the sky than their absolute magnitude (see diagram right).

At last we can refer back to the star colour, temperature and absolute magnitude table from earlier to understand how the amount of gas affects how bright the star truly is. And depending both on how big and bright it is and whether it is near or far away, we may see it brightly, faintly or not at all. Well, with our eyes anyway. Once we dive below around magnitude 6 we need to use things such as binoculars and telescopes to help us view the cosmos. So the next step is to look at what is the best equipment...

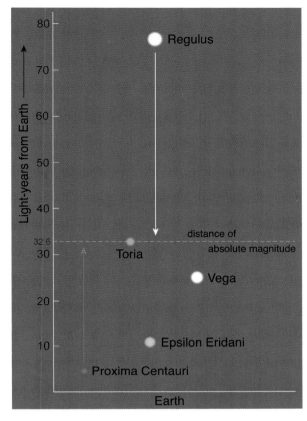

The meaning of absolute magnitude. Stars in the night sky appear to be the same distance away from us when in reality they are not. This means we are unable really to compare one star with another. The magic absolute-magnitude maths stuff arranges all stars (that we can) at a distance of 32.6 light-years away from us, thus forming a stellar identity parade. It's like saying: 'How bright would this star be in the sky if it were moved to this fixed distance?'

Notice Toria – she sits exactly at 32.6 light-years away; therefore she doesn't need to be moved, as her absolute magnitude is the same as her apparent magnitude. A few other choice stars are shown at their actual distances from us: the red arrow indicates that Proxima Centauri would need to move away to get to its absolute magnitude distance – this would make it fainter in the sky and thus its absolute magnitude will be lower than its apparent magnitude. The same thing would happen with Vega and Epsilon Eridani. On the other hand Regulus would have to come closer – it would get brighter and so its absolute magnitude will be higher than its apparent magnitude.

Where Stars Live

On a large scale, stars tend to live in groups, unless some tremendous intergalactic war between the zophids and the cromots has sent a few off into the cosmic backwaters.

The smallest stellar family is the *galactic cluster*. One of these groups typically contains between a dozen and a few thousand stars that were made all together in a nebula within the disc of a galaxy. They appear as a loose assembly of stars, sometimes

Galactic Cluster M7 in Scorpius. (Image courtesy N.A.Sharp, REU program/NOAO/AURA/NSF)

round-ish, sometimes irregular in shape. Over time a galactic cluster will gently disperse, with the stars merging into the general starry background. So, the very fact they are now still a group means they are quite young (astronomically speaking), with timescales only amounting to tens to hundreds of millions of years. Size-wise, they can reach around 60 light-years across.

Next up are *globular clusters*. Unlike the galactic variety, globulars are held together by gravity, so they always stay together. The reason is the vast numbers of stars – from around 10,000 to over one million. Also unlike galactic clusters, globulars are not found in galaxy discs, but spread out in a halo anywhere around the galaxy. This is because they were made

Globular Cluster M10 in Ophiuchus. (Image courtesy NOAO/AURA/NSF)

before the galaxy became disc-shaped, making them very old members of the Universe indeed. All large galaxies seem to have a family of globulars, and as for numbers – well, we have found over 150 (so far) that orbit our Galaxy. These objects can be up to around 300 light-years across.

A *galaxy* is the single largest object of stars in the Universe. In general, one galaxy has between a few million and a few trillion stars – with (depending on the variety) some/lots of/no dust, gas, black holes, etc. The main types are:

Spiral Galaxy NGC 2997 in Antlia. (Image courtesy ESO)

- *Spiral* galaxies – big, flat, round discs with 'arms' of stars and gas spiralling out from a central bulge or bar (as in the case of so-called barred-spiral galaxies)
- *Elliptical* galaxies – rugby-ball-shaped masses of older stars with no cloudy nebulae around to make new stars
- *Lenticular* galaxies – a good mixture of the last two, with the disc-shape of a spiral galaxy (but no actual spiral) and the lack of star-producing gas of an elliptical, making the stars here very old too
- *Irregular* galaxies – those that have been gravitationally affected by a close encounter with a neighbour, so it's difficult to know which one of the former three types a particular irregular galaxy started out as – if it was like one of them at all

Elliptical Galaxy NGC 1316 in Fornax. (Image courtesy ESO)

Lenticular Galaxy NGC 5866 in Draco. (It may be that NGC 5866 is also M102 but we can't be sure – historical references are sketchy, but they do hint at it.) (Image courtesy NOAO/AURA/NSF)

Irregular Galaxy NGC 1427A, with a yellowish little spirally galactic friend (top left) in Fornax. (Image courtesy ESA/NASA/AURA/STScI)

Galaxies range in size from a few thousand to several hundred thousand light-years across.

The Closest and Brightest Stellar Marvels

Here are the closest and brightest stars. Interestingly, note that many of our close stars are very faint (invisible, in fact, to the eye), while many of the brightest stars are very far away.

Nearest Stars	Distance (light-years)	Apparent Magnitude	Constellation
The Sun	very close	-26.72	
Proxima Centauri	4.27	11.05	Centaurus
Rigel Kentaurus A	4.35	0.00	Centaurus
Rigel Kentaurus B	4.35	1.36	Centaurus
Barnard's Star	6.0	9.54	Ophiuchus
Wolf 359	7.8	13.45	Leo
Lalande 21185	8.3	7.49	Ursa Major
Sirius A	8.6	-1.46	Canis Major
Sirius B	8.6	8.44	Canis Major
UV Ceti A	8.7	12.56	Cetus
UV Ceti B	8.7	12.52	Cetus
Ross 154	9.7	10.43	Sagittarius
Ross 248	10.3	12.29	Andromeda
Epsilon Eridani	10.5	3.73	Eridanus
Lacaille 9352	10.7	7.34	Piscis Austrinus

Brightest Stars	Distance (light-years)	Apparent Magnitude	Constellation
Sirius	8.6	-1.46	Canis Major
Canopus	313	-0.72	Carina
Arcturus	37	-0.04	Boötes
Rigel Kentaurus	4.35	0.00	Centaurus
Vega	25	0.03	Lyra
Capella	42	0.08	Auriga
Rigel	773	0.12	Orion
Procyon	11	0.38	Canis Minor
Achernar	144	0.46	Eridanus
Betelgeuse	427	var. 0.58	Orion
Hadar	320	var. 0.6	Centaurus
Altair	16	0.77	Aquila
Aldebaran	65	var. 0.85	Taurus
Antares	604	0.96	Scorpius
Spica	220	0.98	Virgo

CHAPTER 5

Binoculars and Telescopes

When you tell people you are an amateur astronomer, they quite often ask, 'What sort of telescope have you got?' The answer of an f/10, 12-inch Schmidt-Cassegrain will either send them into a frenzied state of excitement or reduce them to total silence before they move on to the next subject. You may as well use this as a default answer anyway, because if you don't have a telescope you'll more than likely only get a quizzical 'call yourself an astronomer?' look. There is just an assumption that if this is your hobby, then *of course* you have a telescope.

The fact that a telescope is not necessary to see many wondrous objects in the starry skies will not cut the mustard with those who have not bought this book, yet have a vague interest in the subject. It is such a shame, as there is a wide variety of amazing objects, events and glows to take a look at most nights of the year. An unaided-eye evening may include some of the following: constellation spotting, hunting for double and coloured stars, variable star observations and seeing a few star clusters, a nebula, a galaxy, some planets, an asteroid, possibly a comet, a couple of shooting stars, a plethora of satellites (including the International Space Station), a bit of zodiacal light, one gegenschein, some noctilucent cloud, a shimmer of aurora, the Milky Way and a crescent Moon.

Having pushed unaided-eye stargazing enough, it has to be said that what telescopes and binoculars do is really open up the vast Universe to 'little old us' sitting in a damp field somewhere just outside Croydon. With their lenses, mirrors and greater light-grabbing capabilities they will both magnify images and enable us to see fainter and further objects than our eye can manage.

Out on a crystal-clear moonless night, with dark adaption (the process of letting the eye get used to the dark) a mere distant memory, the pupils of our eyes are capable of opening to a maximum of about 7mm. This makes the aperture size of our own personal 'body telescope' 7mm, and always on call with a fixed magnification of 1. So there would be little point in making a telescope of 7mm, as the view would be the same as that with the unaided eye. Now, just a small well-constructed telescope of 60mm has more than 70 times the light-grasping power of the eye, and a telescope of 200mm has more than 815 times the power. It's with all this extra light available that the various lenses and mirrors can do their magnifying thing to bring the splendidness of the Universe down to your, your friend's or maybe even your friend's cat's eye.

First Steps to Buying

Let's imagine you're on a quest to buy a telescope: what hurdles will you encounter? Firstly, and maybe surprisingly, there will be the established amateur astronomers to contend with. Some are so 'know it all' that they will tell you not to buy a telescope until you know your way round the night sky with your eyes. This is, quite frankly, bonkers, as any telescopic view of Saturn, for example, has the wondrous ability to encourage you to continue with your night-sky mission. So, the more you can add in to the stargazing mix, the better.

What's more important is getting to know what sort of scope you need and have the money for. While this fun research continues, what can help you to peer quite satisfactorily into the depths are the fantastically portable and often bypassed binoculars. Don't underestimate their usefulness.

Any old binoculars, provided they haven't been knocked about too much or chewed by the dog, will allow you to view some amazing space things, like Jupiter and its four main moons or a visiting comet. You can really see some great sights with something you might have had all along in a dusty cupboard. In this respect you may not have to pay anything for a bit of kit that you can use for your starry pursuits.

Down to binocular details

Binoculars are usually described using numbers, like 7 × 50, said as 'seven by fifty'. The first number, 7 in this case, is the magnification, while the second number, 50, tells you the diameter of the big lenses at the end in millimetres.

A pair of 7 × 50s is actually quite a good basic starting point – although lesser-powered varieties (such as 5 × 20) will do some sort of wizardry with the sky as well. As long as they are not plastic, they didn't come with a child's adventure kit and they look and feel okay, then you should get outside now to try them out.

As the sizes of binoculars grow (70mm, 80mm, etc.) so, as we have seen, does the light-grasping capability and hence the fainter

A few sights for binocular people (not to scale). From left to right, weaving up and down as we go: the Moon, a double star (this one is Epsilon Pegasi), Jupiter and its moons, the crescent phase of Venus, the Pleiades star cluster and the Orion nebula.

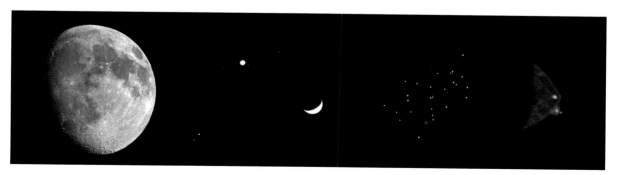

the objects they'll let you see. Two issues arise, though, if you are considering buying larger binoculars: cost and weight. The higher price is probably not worth it unless you are a serious binocularologist – the money you would spend on a really good pair is probably better saved for a telescope – though I realize that that comment will cause uproar in the binocular appreciation society.

As for weight, larger pairs can cause seriously aching arms. If you're a weight-lifting fan, then this is fine, otherwise find a fence or wall to lean on, or observe from a deck chair. This isn't a bad idea with any binoculars, in fact, as after any length of time holding them up can cause some fatigue. The other point about the after-effects of weight is wobbliness. It just ends up being so difficult to hold a pair steady that one of the best solutions is to fit them to a tripod.

I always take a lightweight pair of binoculars out observing with me, even if I'm using a telescope, as I find them great for a general, fast, wide-field sweep – the sky is so full of star clusters and smudgy nebulae that there is always something to discover.

In any case, good binoculars are much better than a cheap, small telescope, so it's worth having a pair around whatever.

Try before you buy

Before you spend any money on binoculars or a telescope, try them out for size. One of the best places to go is your local astronomical society. Here you'll find a varied array of dedicated people with a varied array of equipment – and

experience. Most societies have an observing evening once in a while where you will be able to see telescopes and binoculars in action. However, it is important to remember that nearly everyone, it seems, has a different 'need'. Some want to take pictures, others want a wide-field view, whilst others are interested in detailed planetary work. There may be cost considerations, or perhaps you just want to be able to stick one on a table during a party so everyone can have a look at the Moon.

Maybe you've already got some equipment, but are not sure how to use it. Again an astronomical society will be able to help you out. They really are a good source of information.

The other place to go is a proper telescope shop – and they all sell binoculars too. Here knowledgeable staff are good at working out your best options, balancing price, portability and what you want to do with the thing. As far as anything else goes, I would say that if a shop sells soap and microwaves as well as telescopes, you shouldn't expect too much useful astronomical advice before or after you've bought it.

Down to telescope details

Right back at the top of the chapter there was a 'What sort of telescope have you got?' question. Now that you are prepared for that, the one that follows is invariably, 'What is its magnification?' I guess this totally incorrect idea about how big you can make something appear being the key property that defines a telescope is not altogether surprising – I mean, what else is a telescope for other than to make things look bigger?

Well, actually, some incredible telescopic sights, like delicately structured nebulae and jewelled star clusters, are best seen using *low* powers of magnification – 25 times (written as 25x), for example. Any magnification number given is simply compared to the eye, so viewing something at 25x would bring it 25 times closer to you or, if you prefer, would make it 25 times bigger .

Accordingly, Useful Point Number 1: **Do not even consider magnification when buying a telescope.** Just leave it alone and it'll go away. It means nothing whatsoever. End of.

Of course you do want to magnify things, but any telescope that is worth buying will do that anyway.

Hence, Useful Point Number 2: **How much you can magnify is a consequence of a telescope's aperture – that is, the size of the lens or mirror.** This is, therefore, the biggest consideration. Telescopes get more expensive as their aperture increases. Again we come back to size versus budget, for if you want an easily portable item that you can set up in a matter of moments then you're probably looking for a smallish scope anyway. But for something good that will last a lifetime, go for the biggest you can get!

Worry profusely when telescopes use high-magnification (or *any* magnification) as one of their selling points. I have seen small refracting telescopes announcing: 'With a WHOPPING Two Zillion Times Magnification you're almost in Space!' – avoid these. For a start the optics in these sorts of scopes are not of premium standard, hence the lower cost. If you look down the big lens, often there is a 'stop' (a circular disc) a short way down the tube, which tries to eliminate the imperfections of the image created around the edges of the main, cheaply made objective lens. What the stop also does, not surprisingly, is stop some of the all-important light getting in. So, in reality, what is described as a 60mm refractor may actually be less than a 50mm scope. In fairness, I am not saying these telescopes are totally useless, but you will find them quite limiting, especially if you are a keen amateur. I should know, I had one and I had years of perfectly adequate observing – until I started using well-built scopes and realized what I had been missing.

Even further down the scale, at the very bottom of the telescope barrel, are the cheap plastic varieties that are part of some 'explorer's kit'. For budding astronomers everywhere I would say avoid buying these as you would avoid eating a bowl of mashed peas covered in custard – they are only acceptable as a joke.

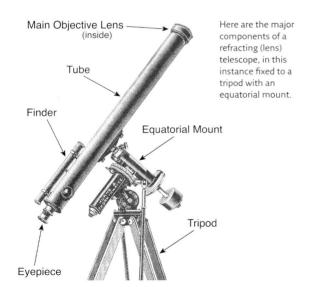

Main Objective Lens (inside)

Tube

Finder

Equatorial Mount

Eyepiece

Tripod

Here are the major components of a refracting (lens) telescope, in this instance fixed to a tripod with an equatorial mount.

Magnification

A telescope collects light and focuses it somewhere – the distance from the lens to this focus, or from the mirror to the focus is known as the *focal length of the telescope*. What you do with the light after that is up to you. For general stargazing you'll more than likely stick in an eyepiece, and it is this little cylinder of lenses that does the magnifying. Eyepieces have numbers on them, and these are the *focal length of the eyepiece*. To work out how much magnification you're going to be using, divide the focal length of the telescope by that of the eyepiece. So, a 900mm telescope with an 18mm eyepiece gives (900/18 =) 50 × magnification.

Lots of people have different ideas about the highest magnification you can use with a telescope. I have used a very conservative 2x the telescope aperture in the table below.

A variety of 'good' eyepieces is what you need to get the best out of your telescope. The item on the right is a Barlow lens – a good one is a nifty piece of kit that can double or triple the magnification of any eyepiece.

Telescope aperture		Highest usable magnification	Faintest objects visible (apparent mag.)	Resolving power (arc seconds)
mm	inches			
60	2.4	120	11.6	1.9"
80	3.1	160	12.2	1.5"
90	3.5	180	12.5	1.3"
100	3.9	200	12.7	1.2"
120	4.7	240	13.1	1.0"
150	5.9	(300)	13.6	0.8"
200	7.9	(400)	14.2	0.6"
250	9.8	(500)	14.7	0.5"

Tying it all together, here is a basic guide to the size of a telescope together with its highest usable power (magnification), the faintest stars you can see with that aperture and the closest two stars can be together and still be seen as two stars (the *resolving power* of the telescope): I've bracketed magnifications above 300 because, even though they are the mathematical highest usable figures, when you get past here you're magnifying any turbulence in the Earth's atmosphere to the detriment of the object you're looking at.

Of course some rules are there to be broken, and I threw caution to the wind one night and happily used 300× magnification on an 80mm telescope – incredibly daring, I know. However, what you get with over-magnification is a loss of brightness, contrast and detail. So break these rules wisely, space fans.

From the table opposite, it is clear that the larger the aperture, the fainter the stars your telescope will allow you to see – the figures refer to perfect conditions with dark-adapted eyes. Some examples of faintness include Pluto, which very few astronomers have seen, as its highest magnitude is 13.8. For this you'll need a telescope of at least 170mm, which would be a very respectable amateur scope indeed. Now take our newly appointed dwarf planet Eris at 19th magnitude: it requires an instrument with an aperture nearing 2m! How serious are you? I've just checked the price if you are interested – approximately £5 million (carry-case not included), but it does come with a CCD camera.

The final column shows the telescope's resolving power. Out there are some great examples of double stars – stars that are visually close for whatever reason – maybe of different colours, and the closer they are the larger the telescope you'll need to 'resolve' them, that is, to see them separately. The measuring units here are *arc seconds*, and these, together with all the other measuring stuff, are the topic of Chapter 6.

Telescope types

Let's just have a look at the different type of telescopes you can choose from. There are three basic designs: Refractors, Newtonian Reflectors and a mixture of the two known as Catadioptrics.

The *refractor* is the well-known, classic-looking, two-lens telescope. The lens at the big end – known as the *objective* – collects the light and focuses it down the tube where a smaller lens, the eyepiece, magnifies the image and throws it into your eye.

Advantages

- Tends to be more robust than a reflector. If you drop it, kick it or it gets trodden on by an elephant (a baby one) it still seems to work. So, apart from a wipe down with a clean cloth every once in a while, it is pretty much maintenance-free.
- Images are high contrast with good definition as there is nothing at all in the light path (see reflectors and Schmidt-Cassegrains, below).
- A good choice for the Moon, planetary and double-star observing.

Eyepiece

EYE

Focus

Lens

Light from some fuzzy blob enters here

ABOVE: The light path in a refractor telescope.

(Image courtesy Paul Wootton, www.graphicnet. co.uk)

LEFT: The objective is the lens, shown here, that captures the starry light; this then enters a refracting telescope and focuses it down at the far end, where you should have plugged in an eyepiece and be observing wonderful things.

The *reflector* opposite does just what it says on the box. It's another classic design with a big curved mirror at the bottom of the tube that collects the light and reflects it back up the way it came. The image then meets another mirror that bounces the image out of the tube to a focus where the eyepiece takes over.

Disadvantages

- More expensive, size for size, than reflectors or Schmidt-Cassegrains.
- Steady mountings are a problem in the cheaper models.
- Only those up to 100mm are easily portable.
- Refractor lenses, especially in the cheaper models, suffer from 'chromatic aberration', where stars have a faint halo of colour around them.

Advantages

- Larger apertures are generally more compact and portable than refractors.
- Good for galaxies, nebulae and star clusters.
- Cheaper, size for size, than refractors.

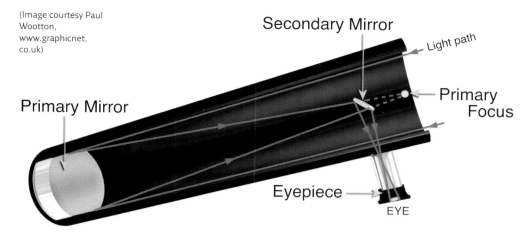

How light bounces around inside a classic Newtonian reflecting telescope.

(Image courtesy Paul Wootton, www.graphicnet.co.uk)

Secondary Mirror

Light path

Primary Mirror

Primary Focus

Eyepiece

EYE

Disadvantages

- They don't care much for the odd bang, which can send mirrors out of alignment.
- Mirrors will slowly lose their reflectivity over time as they lose their coatings due to general chemical life stuff. 'Re-coating' is therefore essential to keep the scope working.
- All mirrors suffer from 'coma', whereby stars look triangular around the edge of the field of view.
- Alignment issues and mirror re-coating mean that reflectors need more maintenance than refractors.

The *catadioptric* types (right) are more modern and use both lenses and mirrors to get the job done. The two main designs are the Schmidt-Cassegrain and the Maksutov-Cassegrain. The Schmidt-Cassegrain is more popular, and lighter.

Advantages

- Very portable due to their compact design.
- Well suited to astro-photography.
- The design virtually removes the refractor's

BELOW: The two versions of the Schmidt-Cassegrain (*top*) and Maksutov.

They do basically the same job, just in different ways.

(Image courtesy Paul Wootton, www.graphicnet.co.uk)

Primary Mirror

Eyepiece

Focus

Corrector Plate

Secondary Mirror

Light

Primary Mirror

Eyepiece

Focus

Meniscus Lens

Secondary Spot

Light

'chromatic aberration' and the reflector's 'coma' disadvantages from any view.

- The best all-round telescope.

Disadvantages

- May be too expensive for most beginners.
- The combination of lenses and mirrors means that a fair amount of light is lost before it reaches the eye, hence images are not as bright as they would be with a refractor or reflector of the same size.

An equatorial mount attached to my refracting telescope in the garden.

Wobble

Next we arrive at something that is probably more important than the telescope itself. Get this wrong and you will not enjoy observing and the telescope will not perform to its best. What am I talking about? It's the Wibbly-Wobbly Factor. It means: how stable is the telescope when fixed to the tripod?

Some scopes are provided with little more than what can only be described as a jelly-mounting system. Without a steady tripod and mount you will only have hours of 'entertainment', watching planets and the like wobbling back and forth across your field of view.

Therefore... Useful Point Number 3: **Make sure the mount of any telescope you buy feels firm and solid.** The tube must not shake if you are to observe successfully, so I would give any potential scope a good (gentle) knock – any image judder should stop within a few seconds. Just to reiterate, this is mainly a problem for cheaper refractors, but watch out for some cheaper reflectors too.

Mounts themselves come in a variety of forms:

The *altazimuth* is the basic mount that comes with small refractors. It just allows the telescope to move up/down, in altitude, and left/right, known as azimuth. Someone put the two together and came up with the name altazimuth. All but the very cheapest include a hand-turned fine-adjustment control connected to the altitude axis. As long as it is well made, this mount is perfectly adequate for observing; if you want to try astrophotography then you need something more sophisticated.

Here I am at Herstmonceux, once home to the Royal Observatory, Greenwich, with an 8-inch (200mm) Newtonian reflecting telescope sitting on a Dobsonian mount.

The *Dobsonian* is a cheap and cheerful low-to-the-ground version of the altazimuth that permits a similar range of movement and is easy to set up and use. It's designed not for refractors but for bigger Newtonian reflecting telescopes. It's also easy to make, so good for DIY telescope enthusiasts, but again not suited to photography.

The *Equatorial* mount is the choice of most established amateurs due to its wide range of uses. Whilst all the crazy RA and dec. markings may seem a bit overwhelming to start with, once you've learnt the Pole Star aligning stuff it becomes second nature. Because you align one of the equatorials axes with that of the Earth, and can fix on a motor drive, the telescope can follow the stars using the magic of a battery.

This means we've finally arrived at a mount suited perfectly to astrophotography.

The Tiny Telescope on the Side

And now for something almost as important as the mount: the finder. This is a tiny version of

Finders, like binoculars, will have all the details of their size and magnification printed clearly on the outside. Make sure any telescope comes with one of a good size, or you won't be able to easily find any faint fuzz-blobs (that's nebulae, star clusters etc.).

a refracting telescope that sits on the side of the tube and simply but effectively helps you find what you want to see. The worth of these (when they are aligned properly!) cannot be overstated. Honestly, even the Moon can be tricky to find without one! A good size to look for is at least a 6 × 30, which will provide a bright, wide-field image. Again, a cheaper telescope may come with a smaller 5 × 24 plus a stop(!), mentioned above, which will cut down the light, allowing you to locate only brighter objects.

All-in-All

There are some *minimum requirements for a basic telescope*: the 60mm example from the table on page 34 is really the smallest *refractor* (lens telescope) you should buy. Anything smaller won't give you the astronomical brightness and quality you want. On the *reflecting* (mirror telescope) front, try to go for at least a 100mm size; otherwise the secondary

mirror and its holder will block too much light from entering the scope. I've seen plenty of *catadioptric* (mirror and lens) scopes and as they are built robustly, all sizes are fine, so just go for the largest you can afford – and will use!

What can you see with a small telescope?

Telescopes will enable you to see everything that binoculars can, but in more detail and with a higher magnification. Of course, due to the larger light-grasp there is also much more. So, prepare to marvel at:

- Craters, mountain ranges, fault lines and other features on the *Moon.*
- A *comet* or two hanging around our part of the Solar System.
- Hundreds of *asteroids* that you can not only locate but also track over a number of days.
- The redness, darker markings and polar ice caps of *Mars.*
- The changing moonlike appearances of *Mercury* and *Venus.*
- *Jupiter* revealing a multitude of dark belts and bright zones running around the giant gas planet. Plus, the Great Red Spot and the moons.
- *Saturn,* its amazing rings and moons, including Titan.
- The planets *Uranus* and *Neptune.*
- Many *double* and *variable stars.*
- Deep sky objects, including *nebulae, star clusters* and *galaxies.*

CHAPTER 6

How to Measure the Sky Without a Ruler

The tip of your little finger at arm's length will cover the Moon; that much is true. However, this is more of an interesting fact than a really useful way of measuring the night. With dexterity you could cover different objects or constellations using other bits of your body: your hand, your arm or even your feet if you are fit enough.

But while I am happy to admit that knowing that the Plough is slightly larger than the outstretched hand, again at arm's length, can help your initial discoveries if you're in the northern hemisphere and if there's no one around who knows how big the darn thing actually is, I do think this is as far as it needs to go. Whilst some astronomy guides depict each constellation in relation to the 'hand', I'm not sure this is really of that much use. Generally you find constellations with a star map, and

can hop merrily around from star to star – the Pointers of the Plough to Polaris, then on to Cassiopeia, across Perseus etc.

It isn't until thoughts turn to looking at smaller objects that the measuring aspect really comes into its own. Knowing the scale of tiny things helps us tremendously – especially in understanding what we should expect to see in binoculars or a telescope. Space bits and bobs vary in size, so if you know how one thing appears in your view, then you'll have an idea of whether another is going to look larger or smaller. It is that simple, and that's why it's worth understanding the size stuff.

Just as we saw in chapter 4 with apparent magnitude, which was only concerned with how bright the thing looked in the sky (not how much light it actually made or how far away it was), that is how we deal with sizes. It is the measure of the apparent size or spacing of object(s) in the sky that matters. To be more precise, it is the angular size of something as seen on the Celestial Sphere from where you are looking. It's because the sky is seen as a hemisphere-dome above us that we have to use an angular (or curved) measurement instead of a straight one (see below).

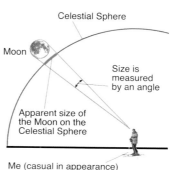

Celestial Sphere

Moon

Size is measured by an angle

Apparent size of the Moon on the Celestial Sphere

Me (casual in appearance)

It's easier for us humans to measure and position anything in space if it is projected onto a dome above our heads – this is the Celestial Sphere. Everything, regardless of its distance or size, can then be brought (or projected) onto this sphere. The green arc shows the appearance of the Moon in the sky (hence, on the sphere), and because it is an arc, and not flat, we measure its size by an angle from where we are looking – astronomers never use tape measures.

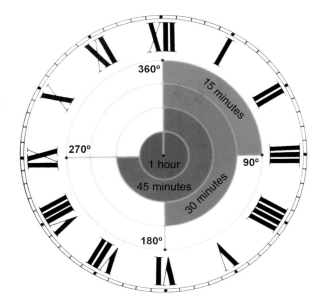

The Space Clock. It's all about the angular movement of the minute hand over time. The outer blue band shows the motion of the minute hand over a period of 15 minutes – during this time it moves through (makes an angular movement of) 90 degrees The greenish half-an-hour band shows that in 30 minutes the hand moves through 180 degrees, and so on.

To demonstrate smaller and smaller angles I have constructed a space clock. Unfortunately, there isn't actually a minute hand sticking out of the Earth and straight up into space, so you need to imagine that bit. But we know that on a regular clock the minute hand goes all the way round in one hour – that's 360 degrees (360°).

In the same way, on our space clock, the imaginary minute hand will sweep out larger angles the more time we give it – 90 degrees in 15 minutes, 180 degrees in 30 minutes and so on (see diagram above).

Now let's take an example. Measuring the Moon, it is found to be just over half a degree across which is large for an astronomical object. But if you look at the space clock, knowing that in 15 minutes of time the hand has covered 90 degrees, you can see that a mere half-degree movement doesn't take very

long. In fact, the minute hand sweeps out half a degree in just five seconds!

If you have a clock, try and see a five-second shift in the position of the minute hand – it isn't easy. One way to make the movement more obvious would be to make the minute hand longer. If it was extended all the way into space (to the Celestial Sphere – which is way beyond the sky), even this slight change would be magnified to the point that it was easily visible. You can get the idea by looking at the picture of myself and the angle of the Moon on page 41. Think of the red lines to the top and bottom of the Moon as the starting and finishing places for the extended minute hand over five seconds.

Just to prove I'm not making it up: if half a degree is covered in five seconds, then one degree is covered in ten seconds. How many ten seconds are there in an hour? 360 of them. Aha! Back to 360 degrees – all the way round.

As one degree is a good chunky solid space size, we'll start from here in our mission to uncover smaller units in an attempt to describe the wondrous Universe of cosmic tiny things. Time to visit the exploded portion of the clock diagram opposite.

On a clock, even though a one-degree angular movement seems tiny, it is divided up into 60 smaller segments known as *arc minutes*, and each one of these is again divided into 60 smaller *arc seconds*.

These units may be written in various forms: *arc minutes* or *minutes of arc* or *arcmin* or just the symbol ′ (e.g. 42′) – they're all the same thing. Following that, the 60 divisions of the arc minute can be: *arc seconds* or *seconds of arc* or *arcsec* or the symbol ″ (e.g. 50″). The diagram of the degree and its divisions again gives some examples of how all this lot works.

With all this knowledge of measuring, take a look at the sizes of some space things in degrees, arc minutes and arc seconds (see box below right).

This table shows that the Sun's average apparent diameter (32′) is one arc minute larger than the Moon's (31′), but to simplify things both of these figures are usually approximated to half a degree. It's an example of a size that can be described by more than one unit – it depends on how specific you want to be.

Not only can we use all this to measure the size of small stuff, we can also measure how some of it moves. Take the constellations: they appear to be fixed to our celestial sphere, unchanging and there forever. But this is not the case. Each star is ploughing its own course through the heavens. Where will life lead us, where are we going and do we have a map? As far as the stars are concerned, yes, more or less, as we can chart where the stars will be in years to come. Of course, because of the enormous distances involved, together with our fleeting existence on this planet, we cannot notice any shift in the main stars ourselves, but generations a long way in the future will eventually not recognize the sky as we see it today.

Symbol
- o degree
- ′ arc minute
- ″ arc seconds

Examples
- $1° = 60′ = 3600″$
- $\frac{1}{4}′\ (0.25′) = 15″$

The degree and its divisions. The minute hand of a clock sweeps out six degrees every minute – shown as one of the yellow segments – which means the hand moves just one degree in ten seconds – this corresponds to one of the six segments shown within the red minute. This may look a small movement, but once you get out into space, each of these one-degree segments is then divided into 60, to give arc minutes, and each of those is divided into 60 again, to give arc seconds.

Space thing	Approximate angular size
Distance from the Pointers in the Plough to Polaris	28°
Length of the Plough	24°
Your outstretched hand at arm's length (roughly)	22°
Distance apart of the Crux Pointers	6°
Distance apart of the Pointers in the Plough	5°
Your forefinger at arm's length	1°
Your little finger at arm's length	¾°
The Sun (average size)	32′
The Moon (average size)	31′
Distance of Ganymede (that's the brightest of the main moons) from Jupiter	6′
Resolution of the unaided eye (this means the ability of your eye to split two objects that are as close together as this)	3′ 25″
Maximum size of Venus	1′
Biggest crater on the Moon	1′
Smallest object your eye can see as a disc and not a point of light	1′
Maximum size of Jupiter	49″
Distance apart of the double star β Cygni (see Cygnus)	34.4″
Smallest object a telescope on Earth can see that has a definite size	1″
Object size limit for the Hubble space telescope in space	0.1″

In the 'Goodbye Plough' diagram, right, you can see that a couple of the stars are moving in different directions to the others – our present-day top-right one, Dubhe, and the one at the end of the Plough's handle, Alkaid. These two are doing their own thing because the other five are part of the Ursa Major Moving Cluster – a scattering of stars made together about 500 million years ago and travelling (though spreading slowly) at about 14km per second through the Galaxy. So, over a period of thousands of years, the shape that inspired us to call this constellation the Plough will evolve into something else entirely.

With the eye we cannot really see stars move in relation to one another, but with a telescope some can be seen to change their positions over just a few years. The two reasons for this are: one, they are fairly 'close' and two, they are bombing along.

This movement of stars in the sky is known as ***proper motion*** and the tiny shifts are measured in arc seconds per year.

The leaders in the proper motion racing stellar stakes are shown in the box below.

So the front runner, Barnard's Star, will move a full arc minute in less than six years, and an easily discernible five arc minutes in under 30. Goodnight!

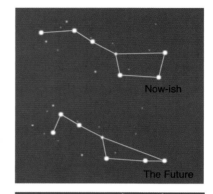

Goodbye Plough, hello Concorde! Will anyone know what Concorde was in 100,000 years when the stars can be joined up in this delta-winged drop-nose aircraft of the twentieth century?

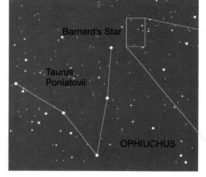

ABOVE: Barnard's Star lies in the constellation of Ophiuchus, near the abandoned constellation of Taurus Poniatovii (catalogued as Anton 1). The time-plot on the right covers 100 years and this movement is equal to around one-half the diameter of the Moon. You're really going to need a telescope if you want to chart its mag. 9.5 flight through the night sky, but it's an interesting project.

Star name	Apparent magnitude	Distance in light-years	Proper motion in arc seconds per year	Constellation
Barnard's Star	9.5	6.0	10.3	Ophiuchus
Kapteyn's Star	8.8	12.7	8.7	Pictor
Groombridge 1830	6.4	30	7.0	Ursa Major
61 Cygni	5.2	11.4	5.2	Cygnus
Lalande 21185	7.5	8.3	4.8	Ursa Major
Wolf 359	13.7	7.7	4.7	Leo
ε Indi	4.7	11.8	4.7	Indus
o₂ Eridani	4.4	16.5	4.1	Eridanus
μ Cassiopeiæ	5.2	24.6	3.8	Cassiopeia
Rigel Kentaurus	-0.01	4.4	3.7	Centaurus
82 Eridani	4.3	19.8	3.1	Eridanus
Arcturus	-0.05	36.7	2.3	Boötes

CHAPTER 7

Starry Charts and the Constellations

Welcome to the night sky. In this chapter there are charts and descriptions of wondrous objects you can see throughout the year wherever you live in the world – or wherever you're going on holiday. Yes, there is no need to be without this book! I also don't want to hear any excuses about light pollution. OK, the fainter objects will be impossible to investigate without nice clear skies, but the constellations, planets, bright nebulae and star clusters are just waiting for you to discover them.

Of course, even though you'll notice the charts are divided hemisphericallywise, depending where you're viewing from, you are not necessarily confined to the northern or southern section just because that is where you happen to live. Once you've started finding your way around the sky – and you'll be amazed how it all comes together with just a few minutes of regular stargazing – you'll soon work out whether you can use the Northern Charts to find Kemble's Cascade, or the Southern Charts to find the Small Magellanic Cloud – or both.

Some of these celestial marvels only require you to gaze in the general direction and there they'll be. Others need a little work with a pair of binoculars, whilst with others, you could do with a powerful telescope that looks like this:

The Gemini North Telescope on Mauna Kea, Hawaii (with the Canada-France-Hawaii Telescope getting in on the act behind). The only reason for showing this picture is that I took it and wanted you to know what a great time I had here.

Starry Symbols

Here are the symbols found on the charts and what they mean:

Symbol	What it is	Mentioned on
⊙	GALACTIC CLUSTER	p. 28
⊕	GLOBULAR CLUSTER	p. 28
▢	NEBULA	p. 16
◇	PLANETARY NEBULA	p. 17
⬭	GALAXY	p. 28

Starry Catalogues

Those M, NGC, B (etc.) letters and abbreviations found on the charts are all catalogues of objects put together by astronomers through the ages. Here's a brief run-down of what's what:

M   Messier Catalogue
A series of objects defined by comet hunter Charles Joseph Messier (1730–1817), dating from 1781.

NGC   New General Catalogue
A collection of catalogues and objects compiled and published in 1888 by Johan Ludvig Emil Dreyer (1852–1926).

IC   Index Catalogue
An extension of the NGC catalogue and also created by Johan Dreyer. The first part (IC 1) appeared in 1895, with the follow-up (IC II) in 1908.

Mel   Melotte Catalogue
Star clusters (galactic and globular) compiled by Philibert Jacques Melotte (1880–1961). It hit the streets in 1915.

B   Barnard Catalogue
A collection of dark nebulae (i.e. ones you can only see because there's something bright behind them) recorded by Edward Emerson Barnard (1857–1923) in 1919.

ANTON Anton Catalogue
Some oddities that have either been forgotten in the mists of time or that deserve more credit than they currently receive in the national press.

Constellation Reference

The 88 western **constellation names** have fine Latin names. The charts given in this chapter identify those containing interesting eye objects, but the full list is given at the end of the chapter.

The three-lettered **abbreviation** is an easy way of identifying a constellation without having to use its full name.

The **possessive** means 'of' or 'belonging to' the constellation and is used for sounding important as if you know what you're talking about – something like, 'Oh, Castor. Of course you mean *alpha Geminorium*.'

Alpha (α) star is the main star of the named constellation – it is not always the brightest and not all constellations have alpha stars with names (or even alpha stars at all for that matter!).

The **magnitude** indicates how bright a star looks in the night sky.

As well as the **colour** of a star pointing out the obvious, it can also tell you how hot it is (see page 23).

The **luminosity** of a star is measured in relation to the Sun. So the Sun, by definition, has a luminosity of 1 and a star with a luminosity of 25 is 25 times that bright.

The **spectral class** refers to a star's mass, temperature and colour, as explained in Chapter 4.

EXAMPLE

LATIN NAME
Monoceros
ENGLISH NAME
The Unicorn
ABBREVIATION
Mon
LATIN
POSSESSIVE
Monocerotis

α STAR
Unicorni
MAGNITUDE
4.1
STAR COLOUR
Yellow-orange

Greek alphabet

α	Alpha	η	Eta	ν	Nu	τ	Tau
β	Beta	θ	Theta	ξ	Xi	υ	Upsilon
γ	Gamma	ι	Iota	ο	Omicron	φ	Phi
δ	Delta	κ	Kappa	π	Pi	χ	Chi
ε	Epsilon	λ	Lambda	ρ	Rho	ψ	Psi
ζ	Zeta	μ	Mu	σ	Sigma	ω	Omega

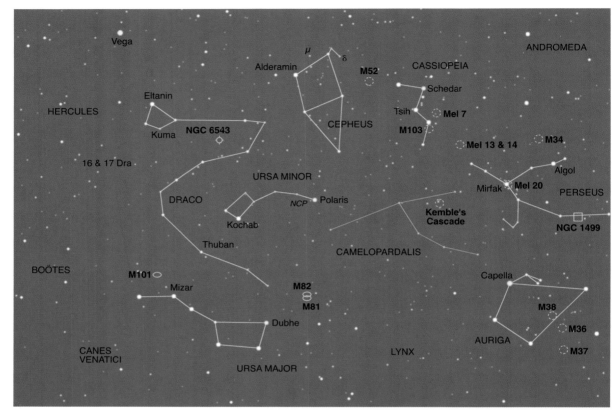

For all the people
and animals
(including polar
bears) living
in or visiting
the northern
hemisphere

The
Northern
Charts

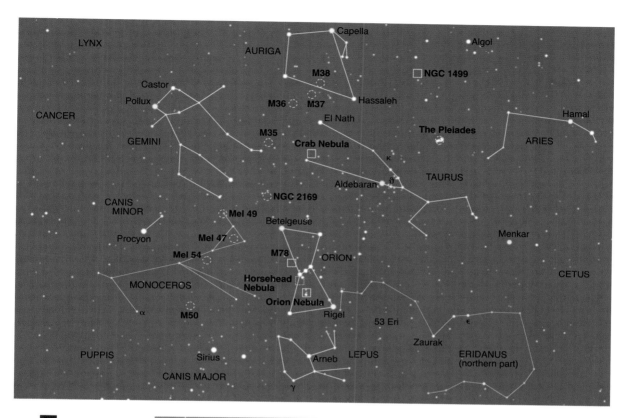

LYNX

AURIGA

Capella

Algol

CANCER

Castor

Pollux

GEMINI

M38

NGC 1499

M36 M37

Hassaleh

El Nath

The Pleiades

Hamal

ARIES

M35

Crab Nebula

κ

TAURUS

CANIS
MINOR

NGC 2169

Aldebaran

ϑ

Procyon

Mel 49

Mel 47

Betelgeuse

Menkar

CETUS

Mel 54

M78

ORION

MONOCEROS

Horsehead
Nebula

Orion Nebula

α

M50

Rigel

53 Eri

ε

Zaurak

PUPPIS

Sirius

Arneb

LEPUS

ERIDANUS
(northern part)

CANIS MAJOR

γ

- ⊙ Galactic Cluster
- ⊕ Globular Cluster
- ☐ Nebula
- ◇ Planetary Nebula
- ⬭ Galaxy

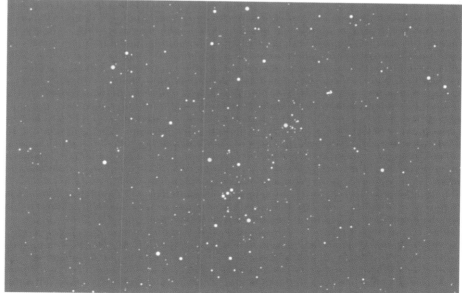

The northern winter
skies looking south.

Orion

LATIN NAME
Orion
ENGLISH NAME
The Hunter
ABBREVIATION
Ori
LATIN
POSSESSIVE
Orionis

α STAR
Betelgeuse
MAGNITUDE
0.3-1.0
STAR COLOUR
Red

The Hunter takes pride of place in the northern hemisphere winter or southern hemisphere summer skies. It has plenty of bright stars: some coloured, like red Betelgeuse and bluish Rigel (although others say it is pure white – make up your own mind); some lined up, like Orion's famous belt or the curving line that forms his shield.

STAR
Betelgeuse
DISTANCE
427 light-years
ABSOLUTE
MAGNITUDE
-7.2 (var.)
LUMINOSITY
15,000 (var.)
SPECTRAL CLASS
M2

STAR
Rigel (β Ori)
DISTANCE
773 light-years
APPARENT
MAGNITUDE
0.12
ABSOLUTE
MAGNITUDE
-8.1
LUMINOSITY
60,000
SPECTRAL CLASS
B8

EYE

DEEP-SKY OBJECT
M42
TYPE
Emission Nebula
MAGNITUDE
4.0
SIZE
1°
DISTANCE IN
LIGHT-YEARS
1,600

The **ORION NEBULA** is a fantastically cloudy work of art that sits in the gap between the 'belt of Orion' and the stars Rigel and Saiph. To get to grips with the size of this thing, imagine the Earth-to-Sun distance as 2.5cm. On that scale, the Orion Nebula would measure nearly 20km across. Impressed? Well,

M42 is actually only the tiny brightest part (lit by the light of new-born stars) of a vast roundy nebula that covers the entire constellation. It truly is gigantinormous! One mystery is that neither Galileo, nor anyone before him, seems to have seen this easy unaided-eye cloud – how do you explain that, dear readers?

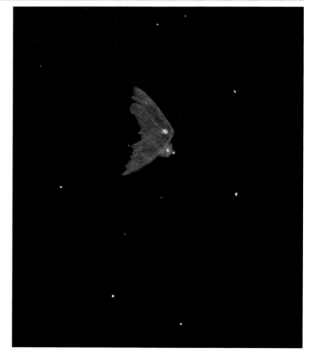

The Orion Nebula as drawn by me on 10 January 1981 using a small 60mm refractor. Peering closely through the eyepiece you can see four stars, known as the Trapezium (θ Ori), sitting in the heart of this cloud. Larger telescopes will show this cloud's delicate structure more clearly.

DEEP-SKY OBJECT
NGC 2169

TYPE
Galactic Cluster

MAGNITUDE
5.9

SIZE
7′

DISTANCE IN
LIGHT-YEARS
3,600

A fine family of around 30 stars sitting in the heart of the faintest visual part of the Milky Way.

The **HORSEHEAD NEBULA**, of course, takes its name from the horsy silhouette shape of dark gas and top-secret space stuff that is blocking the light from more glowing top-secret space stuff further back (called IC 434). This is all really classified, you know.

DEEP-SKY OBJECT
B33

TYPE
Dark Nebula

MAGNITUDE
Not really

SIZE
6′ × 4′

DISTANCE IN
LIGHT-YEARS
1,600

The Horsehead Nebula is not a target for the faint-hearted observer. It is incredibly difficult to see even for expert skywatchers with telescopes that a beginner would describe as 'whoppers'!

DEEP-SKY OBJECT
M78

TYPE
Reflection Nebula

MAGNITUDE
8.3

SIZE
8′ × 6′

DISTANCE IN
LIGHT-YEARS
1,600

You can actually see this in binoculars as a very compact, fuzzy blob, but a telescope is best as it shows the two stars embedded in the cloud that light it up.

A wonderful wide-field view of M78. All the technological paraphernalia that comes with many telescopes today enables you to find objects like this with the press of a button. However, there is nothing like the satisfaction you feel when you have manually star-hopped around an area suddenly to see whatever you're searching for quietly slip into view.

DOUBLE STAR
MINTAKA
δ Ori

MAGNITUDE
2.2 & 6.7

SEPARATION
53″

COLOURS
White & blue

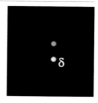

Taurus

LATIN NAME
Taurus
ENGLISH NAME
The Bull
ABBREVIATION
Tau
LATIN
POSSESSIVE
Tauri

α STAR
Aldebaran
MAGNITUDE
0.85
STAR COLOUR
Orange

STAR
Aldebaran
DISTANCE
60 light-years
ABSOLUTE
MAGNITUDE
-0.3
LUMINOSITY
150
SPECTRAL CLASS
K5

Could Taurus be one of the oldest constellations? It depends on how you interpret a cave painting not too far from the entrance to the Lascaux Caves in France. Here a magnificent bull is shown with a small group of stars over its shoulder which looks like (and is in the right position to be) the Pleiades star cluster. There are also spots inside the bull's head that possibly represent the main stars of the constellation.

If this is true, then the estimated age of the painting of around 16,500 years takes our quest to make sense of the heavens back to fascinatingly early times. Alternatively, it could just be an overuse of the imagination – only a time machine will tell.

EYE

DEEP SKY OBJECT
M45
TYPE
Galactic Cluster
MAGNITUDE
1.5
SIZE
1° 50'
DISTANCE IN
LIGHT-YEARS
380

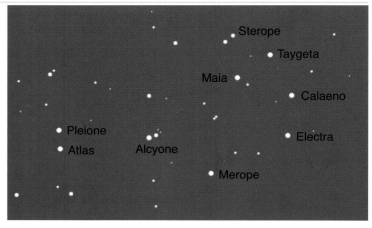

DOUBLE STAR
θ Tau
MAGNITUDE
3.4 & 3.9
SEPARATION
5' 37"
COLOURS
White & yellow

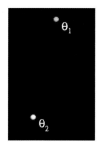

Here is one of the jewels of the skies. How many can you see with the unaided eye? Six stars are very easy, but with super vision you may get to around twelve. The family actually contains many hundreds of stars, and binoculars or a very wide-field view through a telescope will show its glory.

DOUBLE STAR
κ AND 67 TAU

MAGNITUDE
4.2 & 5.7

SEPARATION
5′ 41″

COLOURS
Both white

DOUBLE STAR
σ TAU

MAGNITUDE
4.7 & 5.1

SEPARATION
7′ 18″

COLOURS
Both white

TELESCOPE

M 1
The Crab Nebula

DEEP SKY OBJECT
M1

TYPE
Supernova Remnant

MAGNITUDE
8.4

SIZE
6′ × 4′

DISTANCE IN LIGHT-YEARS
6,300

Welcome on stage – Messier One! Charles had to start his catalogue somewhere, so here it is. The **CRAB NEBULA** is the remains of a star that astronomers saw exploding in July of AD 1054. For 23 days the blast was visible during the daytime, as the brightness reached four times that of the planet Venus. Jumping Jupiter! Great binoculars on a great night can just make out this patch of debris – apparently.

Auriga

LATIN NAME
Auriga
ENGLISH NAME
The Charioteer
ABBREVIATION
Aur
LATIN
POSSESSIVE
Aurigae

α STAR
Capella
MAGNITUDE
0.08
STAR COLOUR
Yellow

STAR
Capella
DISTANCE
42 light-years
ABSOLUTE
MAGNITUDE
0.4
LUMINOSITY
90 + 70
SPECTRAL CLASS
G6 + G2

He's a strange charioteer, this chap as, in all the drawings, he is seen carrying a few goats, and there's no wheeled transportation anywhere. The name of the leading star, Capella, means 'small goat' and the two close stars nearby are 'the kids'. Still no chariot or anything. Maybe it's hidden somewhere in the Milky Way that runs across Auriga – this isn't the brightest part but it's worth a glance in binoculars, as you'll find many star clusters and other stuff.

BINOCULARS

DEEP SKY OBJECT	DEEP SKY OBJECT	DEEP SKY OBJECT
M36	**M37**	**M38**
TYPE	TYPE	TYPE
Galactic Cluster	Galactic Cluster	Galactic Cluster
MAGNITUDE	MAGNITUDE	MAGNITUDE
6.0	5.6	6.4
SIZE	SIZE	SIZE
12′	24′	20′
DISTANCE IN LIGHT-YEARS	DISTANCE IN LIGHT-YEARS	DISTANCE IN LIGHT-YEARS
4,100	4,400	4,200

All three of these clusters (M36, M37 and M38) can be viewed in the same binocular field, but all appear simply as mysterious small smudges. A small telescope will just begin to resolve this one into some of the 60 or so stars that compose it.

This is the largest cluster of the three, made up of around 500 stars.

This cluster looks slightly larger than the other two, but it is definitely the faintest. Some say the stars make a sort of Greek letter pi shape (π) – can't see it myself.

Gemini

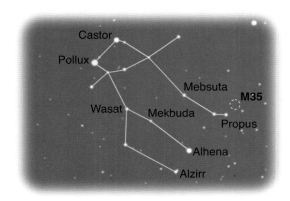

The Greeks saw these two as the twin sons of Leda. It now gets weird – they came out of an egg. Even more strangely (if that's possible), one of the twins, Castor, was the son of King Tyndarus, whilst the other was the son of Zeus. The Romans had a more ordinary story, linking the heavenly Twins with Romulus and Remus, the founders of Rome.

LATIN NAME
Gemini
ENGLISH NAME
The Twins
ABBREVIATION
Gem
LATIN POSSESSIVE
Geminorium

α STAR
Castor
MAGNITUDE
1.58
STAR COLOUR
White

EYE

DEEP SKY OBJECT
M35
TYPE
Galactic Cluster
MAGNITUDE
5.3
SIZE
28'
DISTANCE IN LIGHT-YEARS
2,800

This cluster of around 200 stars may just be seen with the unaided eye on super-crisp nights.

A telescopic view of M35. 1¼ field of view.

STAR
Castor
DISTANCE
49 light-years
ABSOLUTE MAGNITUDE
0.5
LUMINOSITY
58
SPECTRAL CLASS
A1 + A2

STAR
Pollux (β Gem)
DISTANCE
40 light-years
APPARENT MAGNITUDE
1.14
ABSOLUTE MAGNITUDE
0.7
LUMINOSITY
60
SPECTRAL CLASS
K0

This semi-regular star is also a double with a faint ninth-magnitude companion. On 13 March 1781 William Herschel discovered Uranus close to Propus.

BINOCULARS

VARIABLE STAR
PROPUS γ Gem
MAGNITUDE RANGE
3.2 to 4.2
PERIOD
~233 days

Canis Minor

LATIN NAME
Canis Minor
ENGLISH NAME
The Little Dog
ABBREVIATION
Cmi
LATIN POSSESSIVE
Canis Minoris

α Star
Procyon
MAGNITUDE
0.38
STAR COLOUR
Whitish-yellow

The smaller dog of Orion sits off to the left of the Hunter, where the sky is quiet apart from the bright leading star Procyon, meaning 'before the Dog', in reference to that fact it rises before Sirius, the Dog Star (Canis Major). The Greeks were certainly running out of steam by this stage of the 'joining the dots to make a constellation' game – the Little Dog is made of just two stars. Possibly it was based on a small dog that had been run over by a Greek chariot and was thus easy to draw.

Gomeisa

Procyon

STAR	PROCYON
DISTANCE	11 light-years
ABSOLUTE MAGNITUDE	2.6
LUMINOSITY	7
SPECTRAL CLASS	F5

Monoceros

LATIN NAME
Monoceros
ENGLISH NAME
The Unicorn
ABBREVIATION
Mon
LATIN POSSESSIVE
Monocerotis

α Star
Unicorni
MAGNITUDE
4.1
STAR COLOUR
Yellow-orange

Jakob Bartchius, the son-in-law of the famous Johann Kepler, originally named this group Unicorni when he designed it back in 1624. As we know, just a few years later the unicorn became extinct, so it is a pity that the constellation remembering this fine animal is in such a faint area of sky.

Monoceros means 'single horn' – just as rhinoceros means 'nose horn' – well, it's true, you learn something every day!

Mel 49

Mel 47

Mel 54

M50

α

I don't like to use images such as this because most of us don't have a 3.6m telescope sitting on Mauna Kea in Hawaii to play with, but this one just made me put it in. The brighter stars towards the right are NGC 2244, with the central area of the Rosette Nebula sitting all around them and shown in stunning detail. Mouth-watering. (Image courtesy Canada-France-Hawaii Telescope Corporation)

DEEP SKY OBJECT
Mel 47

TYPE
Galactic Cluster

MAGNITUDE
4.8

SIZE
24'

DISTANCE IN
LIGHT-YEARS
5,500

This cluster formed about 4 million years ago from the gas of the surrounding **ROSETTE NEBULA** (NGC 2237) in which it sits. The nebula itself has a magnitude of 6.0 and is 80' × 60' – larger than the size the Moon appears in the sky – but it absolutely needs a telescope. This is a great object if you get into astro-photography. Also catalogued as NGC 2244.

DEEP SKY OBJECT
Mel 49

TYPE
Galactic Cluster with nebula

MAGNITUDE
3.9

SIZE
20'

DISTANCE IN
LIGHT-YEARS
2,400

BINOCULARS

The brightest star here is the fifth magnitude S Monocerotis. It's all in a triangular formation known as the **CHRISTMAS TREE CLUSTER** (see page 23 for a picture). A mighty fine scope on a mountain is needed to see the nearby Cone Nebula which, together with this cluster, is catalogued as NGC 2264.

DEEP SKY OBJECT
Mel 54

TYPE
Galactic Cluster

MAGNITUDE
6.0

SIZE
12'

DISTANCE IN
LIGHT-YEARS
2,800

Binoculars will reveal this cluster as a smudgy sky patch, but a telescope resolves a few dozen stars out of its 80 or so stellar furnaces. It's made even nicer by the string of stars running up and down through the centre. Also catalogued as NGC 2301.

TELESCOPE

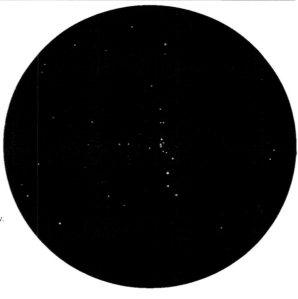

¾° field of view.

DEEP SKY OBJECT
M50

TYPE
Galactic Cluster

MAGNITUDE
5.9

SIZE
16'

DISTANCE IN
LIGHT-YEARS
3,000

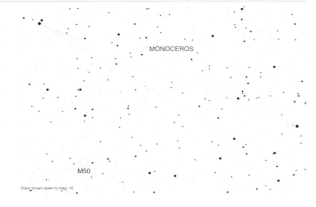

MONOCEROS

M50

Stars shown down to mag. 10

Estimates put the number of stars here in the region of approximately 200, or thereabouts, give or take. Roughly. Lucky M50, it has its own location chart.

Location chart for the star cluster M50 in Monoceros.

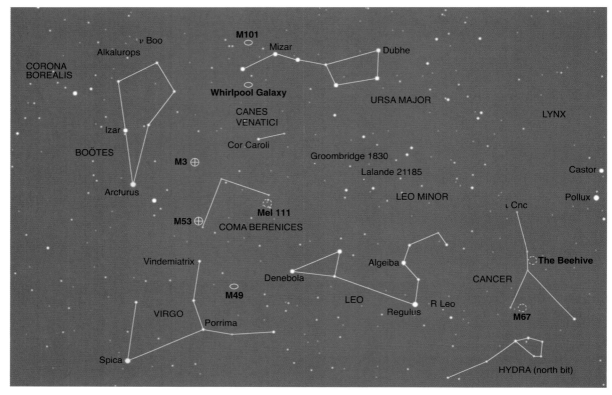

M101

ν Boo

Alkalurops

CORONA
BOREALIS

Mizar

Dubhe

URSA MAJOR

LYNX

Whirlpool Galaxy

CANES
VENATICI

Izar

Cor Caroli

Castor

BOÖTES

M3 ⊕

Groombridge 1830

Lalande 21185

Pollux

ι Cnc

Arcturus

Mel 111

LEO MINOR

The Beehive

M53 ⊕

COMA BERENICES

CANCER

Vindemiatrix

Algeiba

Denebola

M49

LEO

M67

VIRGO

Porrima

Regulus

R Leo

Spica

HYDRA (north bit)

○ Galactic Cluster
⊕ Globular Cluster
□ Nebula
◇ Planetary Nebula
○ Galaxy

The northern spring
skies looking south.

Ursa Major

The Bear was once the lovely lady Callisto. Eventually her son Arcas was changed into a bear and they were both thrown into the sky, presently swinging merrily around the North Celestial Pole as big Ursa Major (mummy) and smaller Ursa Minor (son).

I think she's hungry and she seems to be looking at you.

LATIN NAME
Ursa Major
ENGLISH NAME
The Great Bear
ABBREVIATION
UMa
LATIN POSSESSIVE
Ursae Majoris

α STAR
Dubhe
MAGNITUDE
1.79
STAR COLOUR
Orange

STAR
Dubhe
DISTANCE
125 light-years
ABSOLUTE MAGNITUDE
-1.09
LUMINOSITY
300
SPECTRAL CLASS
K0

STAR
Merak (β UMa)
DISTANCE
80 light-years
APPARENT MAGNITUDE
2.34
ABSOLUTE MAGNITUDE
0.41
LUMINOSITY
69
SPECTRAL CLASS
A1

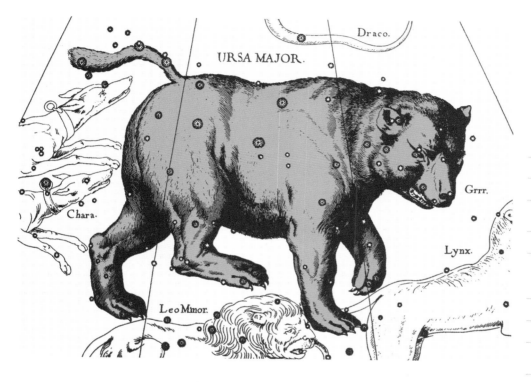

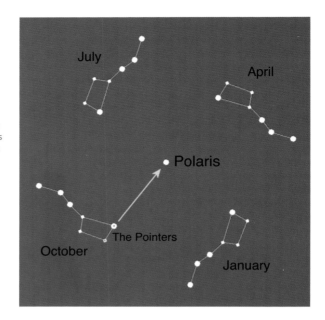

The Plough never sets if you are high enough into the northern hemisphere; it just circles the North Celestial Pole. Here is where you'll find it at around 8pm at certain times through the year. Also, as you can see, the two right-handmost stars of the Plough can be used to find Polaris (the North or Pole Star), which leads to them being known as 'the Pointers'.

The most famous part of Ursa Major is the easily recognizable group of seven stars that form the back and tail of the bear, known as the Plough in the UK, the Big Dipper in the US, der Großer Wagen (the big wagon) in Germany and le Casserole in France, to name a few. The technical term for such a group of stars that make up a recognizable form is an *asterism*, and because this one is quite bright and always above the horizon in a large part of the northern hemisphere it is a particularly useful starting point for finding your way around the sky.

The brightest star in the constellation is Alioth (in the handle of the Plough), which beats the apparent magnitude of Dubhe by just 0.03. This difference is allegedly too small to be discernible to the eye – don't you just love a challenge?

EYE

DOUBLE STAR
MIZAR AND ALCOR
ζ and 80 UMa

MAGNITUDE
2.2 & 4.0

SEPARATION
11' 49"

COLOURS
Both white

These two form the famous optical double star in the bend of the Plough's handle. If you have a telescope you'll see that Mizar is itself a double star, this time a true binary, with the companion 14.4' away. The unmarked star is of magnitude 8.8 and was discovered in 1772 by astronomer Johann Georg Liebknecht , who grandly called it Sidus Ludovicianum ('Ludwig's Star'). In fact, he believed he'd found a new planet because he thought he saw the thing move, which is why he named it after his king, Landgrave Ludwig of Hessen-Darmstadt.

DEEP SKY OBJECT
M81

TYPE
Spiral Galaxy

MAGNITUDE
6.9

SIZE
21' × 30'

DISTANCE IN
LIGHT-YEARS
12 million

This was discovered by Johann Elert Bode in 1774, hence its common name of **BODE'S GALAXY**. There are enough reports to suggest that M81 can be seen with the unaided eye when the skies are crystally dark, but binoculars will just reveal its smudge. Telescopically this elegant galaxy's oval nature becomes apparent as well, making a fine double-galaxy viewing with the M82 sitting just above.

DEEP SKY OBJECT
M82

TYPE
Irregular Galaxy

MAGNITUDE
8.4

SIZE
9' × 4'

DISTANCE IN
LIGHT-YEARS
12 million

Mr Bode discovered the **CIGAR NEBULA**, named because of its elongated shape, on the same night as he found M81. This galaxy has been in gravitational battle with M81 for many years, and the interactions have left it scarred for life. Just look at the picture – it's a mess. M82 is the galaxy towards the top of the image, with larger M81 below.

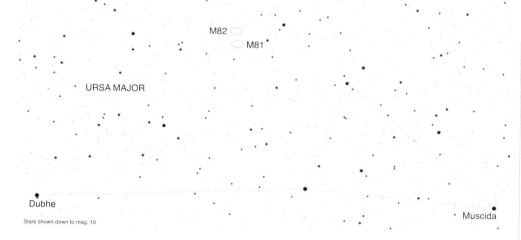

URSA MAJOR

M82

M81

Dubhe

Muscida

Stars shown down to mag. 10

Location chart for the galaxies M81 and M82 in Ursa Major.

DEEP SKY OBJECT
M101

TYPE
Spiral Galaxy

MAGNITUDE
7.9

SIZE
22'

DISTANCE IN
LIGHT-YEARS
27 million

The **PINWHEEL GALAXY** looks like a fairly loose spiral and needs a good dark sky for a small telescope to reveal the brighter central core. Only larger scopes will begin to hint at the spiral arms. Shown on the location chart with M51 (see Canes Venatici page 66).

Ursa Minor

LATIN NAME
Ursa Minor
ENGLISH NAME
The Little Bear
ABBREVIATION
UMi
LATIN
POSSESSIVE
Ursae Minoris

α STAR
Polaris
MAGNITUDE
2.02
STAR COLOUR
Yellow

STAR
Polaris
DISTANCE
~430 light-years
ABSOLUTE
MAGNITUDE
-3.6
LUMINOSITY
2,200
SPECTRAL CLASS
F7

Polaris (α) holds the title of the famous Pole or North Star – at the moment anyway. This is only because it sits more or less directly above the North Pole of the Earth. However, this wasn't always (and won't always be) the case, as the Earth wobbles around over 25,800 years. Mark my words: I reckon that, by the year 3500, the star Errai (γ Cep) will be close enough for it to take the title. Do drop me a line, you know where, to let me know if I'm right.

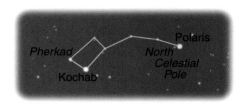

Draco

LATIN NAME
Draco
ENGLISH NAME
The Dragon
ABBREVIATION
Dra
LATIN
POSSESSIVE
Draconis

α STAR
Thuban
MAGNITUDE
3·7
STAR COLOUR
White

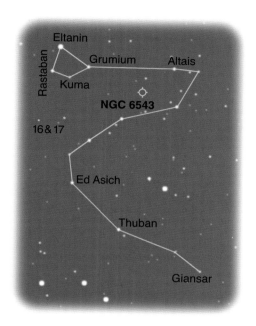

Ladon was the mythical dragon in question. He was guarding the apple tree in the walled Garden of Hesperides, as you do, when Hercules shot an arrow from outside over the wall which killed the dragon. Great shot, but everyone ended up a bit miserable afterwards.

Much has been written about Draco that, if not actually wrong, is verging on the conspiratorially dramatic. Some say its shape has been copied here on Earth by the positioning of ancient temples, which is like me making a dragon shape out of an unusual cloud and writing a book about how aliens came down in cloudy dragon-shaped UFOs to colonize the Earth 20,000 years ago. Also I'm sure I can find an old picture of a stone carving somewhere in Mexico that looks a bit like a dragon and write a bestselling book proving that our ancestors came from Omega Centauri – not a bad idea at all!

Because Thuban used to be the Pole Star, there is no doubt that the ancients would have noted Draco as important, but let's not get carried away into cloud cuckoo land.

BINOCULARS

Here's a tasty double in the head of the Dragon. The two stars are close together, of equal brightness and they stand out well in a star-studded binocular field of view.

DOUBLE STAR
KUMA
ν Dra
MAGNITUDE
4.9 & 4.9
SEPARATION
1′ 02″
COLOURS
Both white

DEEP SKY OBJECT
NGC 6543

TYPE
Planetary Nebula

MAGNITUDE
8.1

SIZE
23' × 18'

DISTANCE IN
LIGHT-YEARS
3,600

This is a super super super (enough!) great bit of star-dying action. Known as the **CAT'S EYE** nebula for its tiny greenyish one-eyed feline-staring appearance in small scopes, it is impressive enough to make it into nearly all the 'best of' catalogues.

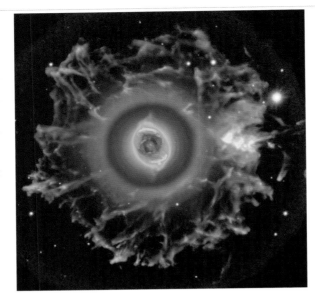

There is absolutely no way in a month of Sundays that you could see all this. It's the most amazing picture from the Nordic Optical Telescope on La Palma in the Canary Islands. The visual bit that our garden telescopes will see is the greenish egg-shape in the middle full of all those red swirls. In this fantastic halo, emission from nitrogen atoms is seen as red, while oxygen atoms are revealed in the green and blue shades. And what is the purpose of showing this? Apart from being an amazing picture, more importantly, it's a perfect example of the fact that there is so much more out there than the eye alone can see. (Image courtesy R. Corradi (Isaac Newton Group), D. Goncalves (Inst. Astrofisica de Canarias))

DOUBLE STAR
16 and 17 Dra

MAGNITUDE
5.5 & 6.4

SEPARATION
1' 30"

COLOURS
Both bluish

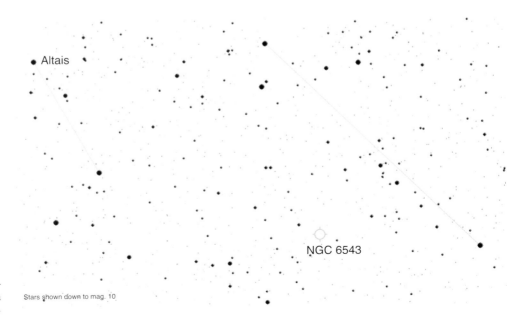

Location chart for the Cat's Eye Nebula (NGC 6543) in Draco.

Stars shown down to mag. 10

Boötes

In history Boötes has apparently changed his outlook on life several times: sometimes protecting the bear Ursa Major and then suddenly becoming a hunter and turning against her. But did he really? The name of the main star, Arcturus, translates as 'guardian of the bear', which makes me lean towards the happier version of the story. So from now on I'm not going to pay any attention to his suggested dark side – maybe he was just having a joke one day, and it got a bit out of hand, you know?

LATIN NAME
Boötes
ENGLISH NAME
The Herdsman
ABBREVIATION
Boo
LATIN POSSESSIVE
Boötis

α STAR
Arcturus
MAGNITUDE
-0.04
STAR COLOUR
Gold

STAR
Arcturus
DISTANCE
37 light-years
ABSOLUTE MAGNITUDE
0.2
LUMINOSITY
115
SPECTRAL CLASS
K1

EYE

Here's a super-wide easy-to-find optical double (that's without light-polluted skies). Of course it's all much easier with binoculars, and armed with these it's just a short hop below to find Alkalurops, another double star.

DOUBLE STAR
ν Boo

MAGNITUDE
5.0 & 5.0
SEPARATION
10' 28"
COLOURS
Deep orange & white

BINOCULARS

Here's a test for your sky and binoculars due to the faint companion. Having said that, my skies are pretty ropy to the northwest, which is where this was when I found it, so don't give me any excuses.

DOUBLE STAR
ALKALUROPS
μ Boo

MAGNITUDE
4.3 & 7.0
SEPARATION
1' 48"
COLOURS
Yellow & cream
(a bit like bananas and custard)

Canes Venatici

LATIN NAME
Canes Venatici
ENGLISH NAME
The Hunting
Dogs
ABBREVIATION
CVn
LATIN
POSSESSIVE
Canum
Venaticorum

α STAR
Cor Caroli
MAGNITUDE
2.9
STAR COLOUR
White

So, I guess one side of the conversation down the pub went something like: 'Ok, Johann "Cheeky" Hevelius, Polish astronomer extraordinaire, you've come up with what constellation this time? Right (reader, now say 'right' again, but this time somewhat slowly and thoughtfully), a dog. Pardon? Two dogs?! Excellent work. Excellent. Out of, um, how many stars? Right (as before), just two stars. But what a great Latin possessive! Cheers!'

EYE

VARIABLE STAR
La Superba
γ CVn
Also TYC 3459-
2147-1 and
HD 110914
MAGNITUDE RANGE
5.2 to 10
PERIOD
251 days

This star was named 'The Superb' by Father Secchi due to what he thought was its amazing red colouring. Note that the star goes below the unaided-eye limit of magnitude 6, so it might be interesting to see at what magnitude it actually disappears and reappears in your location; this will be an indication of how good or bad your skies are. That's if you can see it at all.

BINOCULARS

DEEP SKY OBJECT
M3
TYPE
Globular Cluster
MAGNITUDE
6.2
SIZE
10'
DISTANCE IN
LIGHT-YEARS
39,000

Discovered by Charles Messier in 1764, this is one of the best globular clusters in the skies. Even though you need a larger telescope to start resolving some of its estimated half a million stars, binoculars will reveal it easily, and, due to its 'high' brightness, in very good dark skies, you can just see M3 with your eye.

In 1773 Charles Messier found a faint blob and 'this very faint nebula' entered his famous catalogue. It wasn't until 1845 that M51 became the first 'nebula' to be seen as a spiral by astronomer Lord Rosse (pictured here). The bigger the telescope, the more detail of the spiral arms of what is now called the Whirlpool Galaxy you can make out, but you really need dark skies to see anything at all. Attached to the end of one of its spirals, see if you can spot a smaller fuzzy patch that is another galaxy, NGC 5195.

DEEP SKY OBJECT
M51

TYPE
Spiral Galaxy

MAGNITUDE
8.4

SIZE
11' × 7'

DISTANCE IN LIGHT-YEARS
37 million

Location chart for the Whirlpool Galaxy (M51) in Canes Venatici, and the Pinwheel Galaxy (M101) in Ursa Major.

URSA MAJOR

M101

Mizar

Alkaid

M51

CANES VENATICI

Stars shown down to mag. 10

Coma Berenices

LATIN NAME
Coma Berenices
ENGLISH NAME
Berenice's Hair
ABBREVIATION
Com
LATIN POSSESSIVE
Comae Berenices

Coma Berenices became a fixed constellation when it was catalogued by Tycho Brahe in 1601. Before that not many people thought 'hair' made a substantial addition to the night sky hall of fame, even if these were the locks of Queen Berenice, the wife of Ptolemy III, King of Egypt.

α STAR
Diadem
MAGNITUDE
4.3
STAR COLOUR
Yellow

EYE

DEEP SKY OBJECT
MEL 111
TYPE
Galactic Cluster
MAGNITUDE
2.7
SIZE
4′ 30″
DISTANCE IN LIGHT-YEARS
265

The **COMA STAR CLUSTER** of about 45 stars used to be the fuzzy hair at the end of Leo the Lion's tail; now it forms the flowing hair of Queen Berenice.

15° field of view.

BINOCULARS

DEEP SKY OBJECT
M53
TYPE
Globular Cluster
MAGNITUDE
7.6
SIZE
12′
DISTANCE IN LIGHT-YEARS
58,000

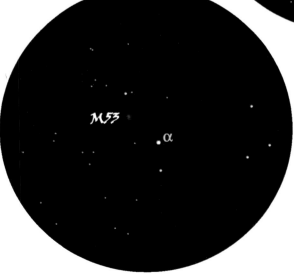

5° field of view.

In 1775 Bode found this distant but fine example of a globular cluster that sits just to the top right of the leading star Diadem. Binocularwise, it won't hit you between the eyes immediately, but do keep going. Telescopically it's slightly oval in appearance.

Leo

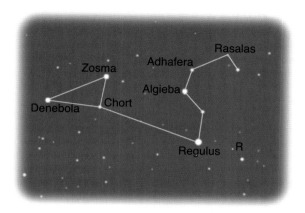

Leo is a very old constellation representing the jolly Greek lion that spent its days eating people and wandering around in the Nemean forest. Its skin, just like a sci-fi android, repelled iron, and who knows what else, so that, for the first of his twelve labours, Hercules had to resort to strangling it.

LATIN NAME
Leo
ENGLISH NAME
The Lion
ABBREVIATION
Leo
LATIN POSSESSIVE
Leonis

α STAR
Regulus
MAGNITUDE
1.35
STAR COLOUR
Bluish

STAR
Regulus
DISTANCE
69 light-years
ABSOLUTE MAGNITUDE
-0.3
LUMINOSITY
134
SPECTRAL CLASS
B7

Regulus also has the names **COR LEONIS** ('Heart of the Lion') and **REX** ('the King').

BINOCULARS

DOUBLE STAR
REGULUS
α Leo
MAGNITUDE
1.35 & 7.7
SEPARATION
2′ 57″
COLOURS
White & orange

DOUBLE STAR
ALGIEBA
γ Leo
MAGNITUDE
2.2 & 3.5
SEPARATION
5″
COLOURS
Orange & yellow

This is a strange, Mira-type (long-period) reddish-purply variable. Catch it when it's bright, or else – it likes to spend most of its time in faint mode.

TELESCOPE

VARIABLE STAR
R LEO
Also TYC 831-521-1 and HD 84748
MAGNITUDE RANGE
4.4 to 11.3
PERIOD
312 days

Cancer

LATIN NAME
Cancer
ENGLISH NAME
The Crab
ABBREVIATION
Cnc
LATIN
POSSESSIVE
Cancri

α STAR
Acubens
MAGNITUDE
4.25
STAR COLOUR
White

This ancient group, built of faint stars between Leo and Gemini, was the crab sent by Hydra to have words with Hercules. Unfortunately, for the crab at least, Hercules trod on it and... crab pâté!

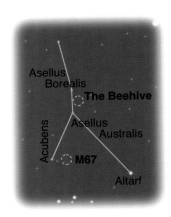

EYE

DEEP SKY OBJECT
M44
TYPE
Galactic Cluster
MAGNITUDE
3.7
SIZE
1° 35'
DISTANCE IN
LIGHT-YEARS
577

Detail of the Beehive seen through a nice pair of binoculars. 5° field of view.

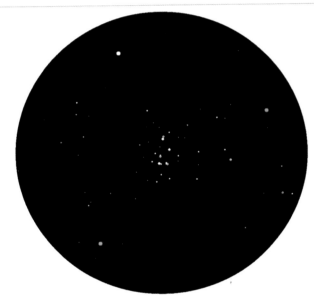

The **BEEHIVE**, or **PRAESEPE** cluster, has hundreds of stars, many of them doubles, visible to the eye as with this ever-popular fuzzy patch. Due to its brightness, M44 has been known from 'way back when'.

BINOCULARS

DEEP SKY OBJECT
M67
TYPE
Galactic Cluster
MAGNITUDE
6.1
SIZE
30'
DISTANCE IN
LIGHT-YEARS
2,700

A very old, weak, loose and slightly sad cluster, not so much a Beehive, rather an old abandoned warehouse where they used to store carrots. An easy target for binoculars, but better with a telescope.

TELESCOPE

DOUBLE STAR
ι **CNC**
MAGNITUDE
4.5 & 4.5
SEPARATION
30"
COLOURS
Yellow & blue

Virgo

The Maiden is an ancient group associated with the Roman goddess of justice as well as the Greek goddess of the harvest. The name of her leading star, Spica, translates as 'ear of wheat', so the Greeks have it, I think.

LATIN NAME
Virgo
ENGLISH NAME
The Maiden
ABBREVIATION
Vir
LATIN POSSESSIVE
Virginis

α STAR
Spica
MAGNITUDE
0.98
STAR COLOUR
Bluish- white

STAR
Spica
DISTANCE
220 light-years
ABSOLUTE MAGNITUDE
-3.2
LUMINOSITY
2,100
SPECTRAL CLASS
B1

BINOCULARS

This is the brightest of over 10,000 galaxies that float about in Virgo and the adjoining constellation of Coma Berenices.

DEEP SKY OBJECT
M49
TYPE
Elliptical Galaxy
MAGNITUDE
8.4
SIZE
9' × 7.5'
DISTANCE IN LIGHT-YEARS
60 million

TELESCOPE

These drawings were made in the mid-1830s and show the great orbital movement of Porrima over just a few years.

DOUBLE STAR
PORRIMA
γ Vir
MAGNITUDE
3.6 & 3.6
SEPARATION
1.8"
COLOURS
Both white

Camelopardalis

LATIN NAME
Camelopardalis
ENGLISH NAME
The Giraffe
ABBREVIATION
Cam
LATIN POSSESSIVE
Camelopardalis

α STAR
α Cam
MAGNITUDE
4.3
STAR COLOUR
Blue

Empty patches of the sky are usually the last left to fill with constellations, and when they are there's no guarantee of success. Custos Messium (named in honour of astronomer Charles Messier) and Rangifer (the Reindeer) were two early contenders, but it was the Giraffe who finally made it onto the starry stage. Well done.

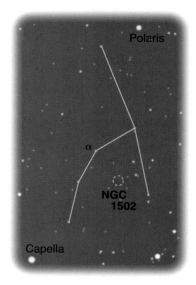

BINOCULARS

DEEP SKY OBJECT
KEMBLE'S CASCADE
TYPE
Asterism
MAGNITUDE
7.0
SIZE
2° 30'

This is a great binocular target, forming a long cascading chain of around 20 stars. It leads off from the small 7' fuzz of sixth magnitude that is NGC 1502, but the Cascade is what you'll be looking at.

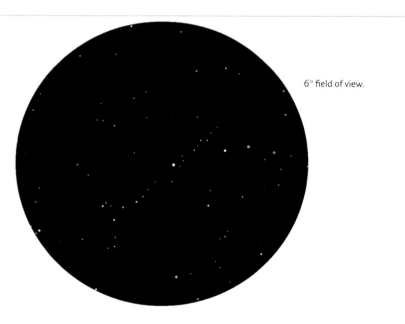

6° field of view.

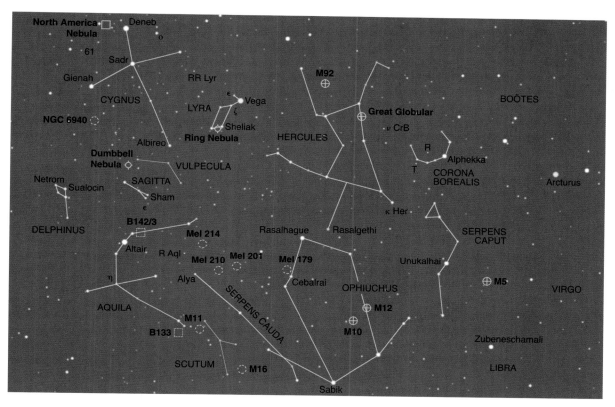

North America Nebula □
Deneb
61
Sadr
Gienah
CYGNUS
NGC 6940 ○
Albireo
Dumbbell Nebula ◇
Netrom
Sualocin
ε
DELPHINUS
B142/3 □
Altair
R Aql
η
AQUILA
B133 □
M11
SCUTUM
M16 ○
o
RR Lyr
LYRA
ε
Vega
ζ
Sheliak ◇
Ring Nebula
VULPECULA
SAGITTA
Sham
Mel 214
Mel 210
Mel 201
SERPENS CAUDA
M92 ⊕
HERCULES
Great Globular ⊕
ν CrB
R
T
κ Her
Rasalhague
Rasalgethi
Mel 179
Cebalrai
OPHIUCHUS
M12 ⊕
M10 ⊕
Sabik
BOÖTES
Alphekka
CORONA BOREALIS
Arcturus
SERPENS CAPUT
Unukalhai
M5 ⊕
VIRGO
Zubeneschamali
LIBRA
Alya

- ○ Galactic Cluster
- ⊕ Globular Cluster
- □ Nebula
- ◇ Planetary Nebula
- ◯ Galaxy

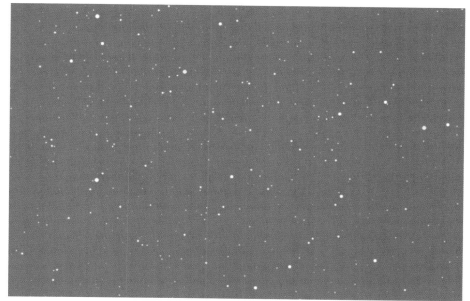

The northern summer skies looking south.

Cygnus

LATIN NAME
Cygnus

ENGLISH NAME
The Swan

ABBREVIATION
Cyg

LATIN POSSESSIVE
Cygni

α STAR
Deneb

MAGNITUDE
1.25

STAR COLOUR
White

Folks, this is a long story of a foolhardy youngster. Condensed for speed, Apollo's son, Phaeton, drove the Sun chariot at such speed that it scorched the Earth. Zeus killed him to stop any more damage, whereupon his body fell into the river. His grieving friend Cygnus (getting there) went to search the river, swimming up and down like a swan...aha!

Phaeton's sisters, the Heliades, also grieved, but they didn't make it into the sky – they were turned into alder trees instead. There's a moral in there somewhere.

STAR
Deneb

DISTANCE
1,500 light-years

ABSOLUTE MAGNITUDE
-7.2

LUMINOSITY
70,000

SPECTRAL CLASS
A2

The bright summer sky stars of Deneb (~2,100 light-years away), Vega (25 light-years) and Altair (16 light-years) form the famous Summer Triangle.

EYE

DEEP SKY OBJECT
M39

TYPE
Galactic Cluster

MAGNITUDE
4.6

SIZE
32′

DISTANCE IN LIGHT-YEARS
825

M39 was placed in the Messier Catalogue in 1764, but because it is so bright Aristotle had noted this starry groupiness as far back as ancient Greek times. Due to general 'life' this cluster is shown in the top right of the northern October-to-December chart.

DEEP SKY OBJECT
NGC 7000

TYPE
Nebula

SIZE
2°

DISTANCE IN LIGHT-YEARS
1,600

The **NORTH AMERICA NEBULA** can 'allegedly' be picked out on really dark nights away from light pollution. This cloudy patch sits in the heart of the Milky Way and it really is incredible how much like its name it looks.

BINOCULARS

DOUBLE STAR
o₁ CYG

MAGNITUDE
3.8 & 7.0

SEPARATION
1′ 41″

COLOURS
Orange & turquoise

A great double as the colours of the stars really do stand out against the background wash of Milky Way stars.

TELESCOPE

DOUBLE STAR
ALBIREO
β Cyg

MAGNITUDE
3.1 & 5.1

SEPARATION
34″

COLOURS
Golden & blue

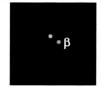

Legally this is the finest double star in the heavens.

Not only a binary system, but moving about all over the place! In fact, in 1792 Giuseppe Piazzi (who discovered the first asteroid, Ceres, in 1801) named it the 'Flying Star'. Then,

DOUBLE STAR
61 CYG

MAGNITUDE
5.2 & 6.0

SEPARATION
29″

COLOURS
Yellow & orange

in 1838, this became the first star to have its distance measured. Using a method called **parallax** (see Chapter 4) Friedrich Wilheim Bessel calculated the distance to 61 Cyg as 11.1 light-years.

Lyra

A constellation known since antiquity, depicting the musical instrument invented by Hermes and given to his half-brother Apollo. Not that musical, Apollo passed it on to his son Orpheus, the chap who made the music for the Argonauts.

Vega (α) is a relatively close (27-light-year-distant) star that held the post of the Pole Star 11,000–12,000 years ago – and it'll do the job once more in about AD 14,500. This is all due to the spinning wobble of the Earth, called precession, that slowly moves our 23.5 degree axial tilt around, drawing out a circle in the sky over our planet's 25,800 year wobble. Therefore, the point above the North or South Pole changes over the same period, hence the changes in Pole Stars.

LATIN NAME
Lyra
ENGLISH NAME
The Harp
ABBREVIATION
Lyr
LATIN POSSESSIVE
Lyrae

α STAR
Vega
MAGNITUDE
0.03
STAR COLOUR
Bluish- white

STAR
Vega
DISTANCE
25 light-years
ABSOLUTE MAGNITUDE
0.6
LUMINOSITY
52
SPECTRAL CLASS
Ao

EYE

DOUBLE STAR
ε¹ and ε² Lyr
MAGNITUDE
5.0 & 5.0
SEPARATION
3.5′
COLOURS
Yellow & orange

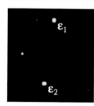

Here's a double star worth taking a look at. These stars are a widely separated optical double which is supposed to be a test of good eyesight. Now pick up your telescope because ε Lyr is also a double-double – two stars that are themselves two stars. What a bonanza of stellar superness: four stars for the price of one!

VARIABLE STAR
SHELIAK
β Lyr
MAGNITUDE RANGE
3.3 to 4.3
PERIOD
12.9 days

An eclipsing binary variable. This is the first of the Beta Lyr-type variables where the stars orbit so close together that gravity pulls them out of shape – making them look more like eggs than balls.

BINOCULARS

DOUBLE STAR
ζ LYR
MAGNITUDE
4.3 & 5.9
SEPARATION
44″
COLOURS
Topaz & greenish

You'll need steady hands, or a wall to rest on, for this double, but it's well worth a

For those of you with the right gear attached to your telescope, this is the sort of thing you can get out of the marvellous Ring Nebula.

TELESCOPE

DEEP SKY OBJECT
M57
TYPE
Planetary Nebula
MAGNITUDE
9.0
SIZE
1.3′ × 1′
DISTANCE IN LIGHT-YEARS
2,300

The **RING NEBULA** appears as an oval fuzzy patch in a small telescope. Larger instruments will show it as a true ring and with colour CCD stuff (if you get into that) the wonderful rainbow colours of this large stellar marvel are easily imaged.

Is it a ghost? Of course not, it's the marvellous M57 through a small telescope.

peek – the yummy binocular view includes Vega and the ε Lyrae double.

VARIABLE STAR
RR LYR
also TYC 3142-494-1 and HD 182989
MAGNITUDE RANGE
7.1 to 8.1
PERIOD
13.6 hours

Sitting very close to the border with Cygnus, this stellar wonder is the originator of an important class of stars used to calculate distances – especially for globular clusters. While pondering that, why not watch it change from white to yellowish over its relatively short period?

Aquila

LATIN NAME
Aquila
ENGLISH NAME
The Eagle
ABBREVIATION
Aql
LATIN
POSSESSIVE
Aquilae

Look out Ganymede, Aquila, the good friend of Zeus, is coming to get you. Too late. Off they fly to Mount Olympus, where Zeus gets Ganymede to serve the drinks – he's a lazy one, that Zeus.

α STAR
Altair
MAGNITUDE
0.77
STAR COLOUR
White

EYE

VARIABLE STAR	A Cepheid variable.
γ AQL	Use the nearby
MAGNITUDE RANGE	β Aql, which shines
3.5 to 4.3	away at magnitude
PERIOD	3.7, as a useful
7.176 days	comparison star.

STAR
Altair
DISTANCE
16 light-years
ABSOLUTE
MAGNITUDE
2.3
LUMINOSITY
10.7
SPECTRAL CLASS
A7

BINOCULARS

DEEP SKY OBJECT
MEL 214
TYPE
Galactic Cluster
MAGNITUDE
6.7
SIZE
13'
DISTANCE IN
LIGHT-YEARS
1,200

There are around 100 stars, including some doubles, in this watercoloury triangular wish-wash of the sky. Nicely, there is a star at each corner of the triangle as well as some good doubles. The bigger your binoculars, the better. Also catalogued as NGC 6709. See Ophiuchus for its location chart.

DEEP SKY OBJECT
B 133
TYPE
Dark Nebula
MAGNITUDE
It's dark
SIZE
9' × 5'
DISTANCE IN
LIGHT-YEARS
1,300

This is quite a small obscuring cloud that is a challenge in anybody's book. I should really have put it under 'Telescope', but with excellent seeing in a dark place you may just find it.

DEEP SKY OBJECT
B 142/143
TYPE
Dark Nebula
MAGNITUDE
It's also dark
SIZE
30'
DISTANCE IN
LIGHT-YEARS
1,300

Together these two nebulae make Barnard's famous 'E' nebula – just take a look and you'll see what I'm going on about. These patches of Milky-Way-obscuring cloud are easier to see (or not, as the case may be – how does one see black?) than the other darkness, B133.

Well that's 'E' covered, how about looking for the other 25 letters of the alphabet in other nebulae, star clusters, galaxies, etc.?

VARIABLE STAR
R AQL
Also TYC 1040-241-1 and HD 177940
MAGNITUDE RANGE
5.5 to 12
PERIOD
284.2 days

You'll need a telescope to see the whole of this red Mira-type star's nine-monthish cycle.

Delphinus

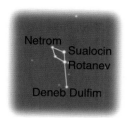

This is the dolphin that helped Poseidon, the god of the sea, to get his wife Amphitrite. Later this watery champ saved Arion, Poseidon's son, when he was being attacked at sea, by distracting the assailant with plenty of dolphiny playfulness using hoops and stuff.

I always like to mention cheeky astronomers and here the cheekiness is hidden in the star names. Alpha Delphini is Sualocin, while beta Delphini is Rotanev. Reverse them and you get Nicolaus Venator, who was an assistant to the Italian astronomer Giuseppe Piazzi. Now that is 'classic'.

LATIN NAME
Delphinus
ENGLISH NAME
The Dolphin
ABBREVIATION
Del
LATIN POSSESSIVE
Delphini

α STAR
Sualocin
MAGNITUDE
3·77
STAR COLOUR
Bluish-white

TELESCOPE

DOUBLE STAR
NETROM
γ Del
MAGNITUDE
4.5 & 5.5
SEPARATION
10″
COLOURS
Green & yellow

Sagitta

An old constellation named by the Romans, Sagitta is a fine little group sitting in the lower part of the Summer Triangle. It simply seems to be one of Sagittarius' arrows that got stuck in the sky after one of those nights. The authorities are apparently trying to play down the incident.

LATIN NAME
Sagitta
ENGLISH NAME
The Arrow
ABBREVIATION
Sge
LATIN POSSESSIVE
Sagittae

VARIABLE STAR
U SGE
Also TYC 1607-913-1 and HD 181182
MAGNITUDE RANGE
6.4 to 9.2
PERIOD
3d 9d 8m

An Algol-type eclipsing binary star sitting off to the right of **ANTON o**, the Coathanger, in Vulpecula. The minimum (the period

of time when the star is at it's faintest) lasts for 1 hour 40 minutes, after which the brightness increases rapidly over just a few minutes.

BINOCULARS

DOUBLE STAR
ε SGE
MAGNITUDE
5.7 & 8.3
SEPARATION
1′ 29″
COLOURS
Orange & yellow

α STAR
Sham
MAGNITUDE
4.37
STAR COLOUR
Yellow

Vulpecula

LATIN NAME
Vulpecula
ENGLISH NAME
The Fox
ABBREVIATION
Vul
LATIN POSSESSIVE
Vulpeculae

α STAR
Anser
MAGNITUDE
4·44
STAR COLOUR
Orange

Sitting immediately under Cygnus, Johann Hevelius originally named this constellation Vulpecula cum Ansere, the Fox and Goose. We've since lost the goose, I guess because the fox got a little peckish. Whatever the story, it is not the most distinctive of groups, but there are some really great binocular objects to find.

Anyone with the digital kit and patience can begin to get some fine pictures of the cosmic wonders out there – here's a great image of M27, the Dumbbell Nebula. (Image courtesy Paul Whitmarsh)

DOUBLE STAR
ANSER
α Vul
MAGNITUDE
4.4 & 5.8
SEPARATION
6' 50"
COLOURS
Red & orange

An optical double with the star 8 Vul.

BINOCULARS

DEEP SKY OBJECT
ANTON 0
TYPE
Asterism
MAGNITUDE
3.6
SIZE
1°

The **COATHANGER**, also known as CR399, or Brocchi's Cluster, takes its name from the astrotastically fab shape of its ten main stars. With the unaided eye in a really good dark sky you can just see it as a fuzzy patch. Formerly thought to be a true grouping of stars, it now seems that only some of the group were formed together and this is more of a chance wardrobe-alignment of stars.

DEEP SKY OBJECT
NGC 6940
TYPE
Galactic Cluster
MAGNITUDE
6.3
SIZE
31'
DISTANCE IN LIGHT-YEARS
We simply don't know!

A good, rich cluster of around 100 stars.

DEEP SKY OBJECT
M27
TYPE
Planetary Nebula
MAGNITUDE
7.6
SIZE
8' × 6'
DISTANCE IN LIGHT-YEARS
1,250

The **DUMBBELL NEBULA** is an easy binocular target, but better with a telescope. It sits in the Milky Way, so there are a myriad of other things to look at in the surrounding area. Well done. Hip hip hooray!

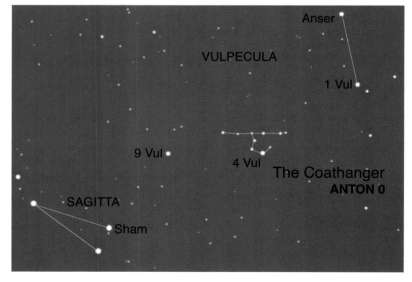

M27, right, gets full marks from me for location, sitting as it does under an 'M' of stars in the northern hemisphere, or a 'W' for southernites.

Hercules

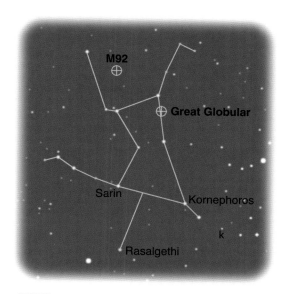

This ancient group is one of the very few constellations in the northern hemisphere whose figure appears upside down. It is associated with Zeus' son who was placed in the sky after finishing twelve impossible 'labours' – how is that possible? Originally, though, he was just some kneeling man who kept his foot on Draco the dragon's head – like that's also possible!

LATIN NAME
Hercules
ENGLISH NAME
Hercules
ABBREVIATION
Her
LATIN POSSESSIVE
Herculis

α STAR
Rasalgethi
MAGNITUDE
3.1-3.9
STAR COLOUR
Red

EYE

DEEP SKY OBJECT
M13
TYPE
Globular Cluster
MAGNITUDE
5.7
SIZE
23′
DISTANCE IN LIGHT-YEARS
25,300

VARIABLE STAR
RASALGETHI α Her
MAGNITUDE RANGE
2.8 to 4.0
PERIOD
~3 months

Some stars really do give value for money – this one is not only variable, but also a double. Rasalgethi's companion has a magnitude of 5.4 with a 5″ separation, which means a telescope is called for.

The **GREAT GLOBULAR**, as it is named, is the finest northern-hemisphere globular cluster. It is best observed from a dark site when its visual 10′ size is fairly easy to see with the unaided eye. This round 'blob' is great in binoculars, but you need a telescope to start resolving the 'fuzz' into stars. We sent a signal in this direction in 1974, inviting aliens to visit. Seeing as the message will not get there until around the year 27,300, then they have to send their attack vessels, which will take another 25,300-ish years to get here, thankfully I won't be around to witness the consequences of our rash actions.

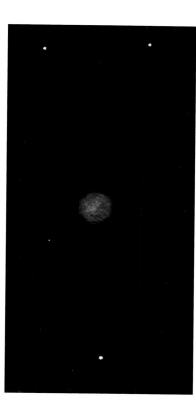

Here's a drawing I made of M13 on 30 July 1981, using a small 60mm refractor.

BINOCULARS

DEEP SKY OBJECT
M92
TYPE
Globular Cluster
MAGNITUDE
6.5
SIZE
11.2
DISTANCE IN LIGHT-YEARS
26,700

It's worth making a comparison between this easy globular and M13 (mentioned above). Johann Bode found this gem in 1777.

TELESCOPE

DOUBLE STAR
κ HER
MAGNITUDE
5.3 & 6.5
SEPARATION
28.4″
COLOURS
Both orange

Ophiuchus

LATIN NAME
Ophiuchus
ENGLISH NAME
The Serpent
Bearer
ABBREVIATION
Oph
LATIN
POSSESSIVE
Ophiuchi

α STAR
Rasalhague
MAGNITUDE
2.08
STAR COLOUR
White

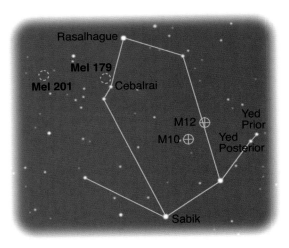

I prefer this constellation's old name, Serpentarius, which describes his serpent-bearing qualities very nicely. No one knows the real story, but he could be Aesculapius, the Greek god of medicine – although he couldn't practise much as he carried a serpent wherever he went and his patients got a little jittery. In his hands the serpent stretches to the right as the constellation of Serpens Caput, its head, and to the left as Serpens Cauda, its tail.

Here we find Barnard's Star (Velox Barnardi), named after its discoverer Edward Barnard. This is the third-closest star to us, at just six light-years away, but the most interesting fact is that it has the largest proper motion of any star, moving at an incredible 10.25 arc minutes per year. Unfortunately Barnard's Star is truly faint, and even its closeness only helps the magnitude reach 9.5. Its location chart, showing a 100-year movement, is in on page 44.

DEEP SKY OBJECT
M10
TYPE
Globular Cluster
MAGNITUDE
6.6
SIZE
14.5'
DISTANCE IN
LIGHT-YEARS
14,400

DEEP SKY OBJECT
M12
TYPE
Globular Cluster
MAGNITUDE
6.6
SIZE
15.1'
DISTANCE IN
LIGHT-YEARS
16,000

M10 sits quite close to M12, so it's easy to compare them – you'll find this is the smaller and neater of the two.

M12, together with its nearby friend M10, are the finest examples of more than 20 globular clusters that find themselves in Ophiuchus.

Compared to M10, this one has a slightly larger appearance. A telescope reveals it to be a less compact object with overall fainter stars.

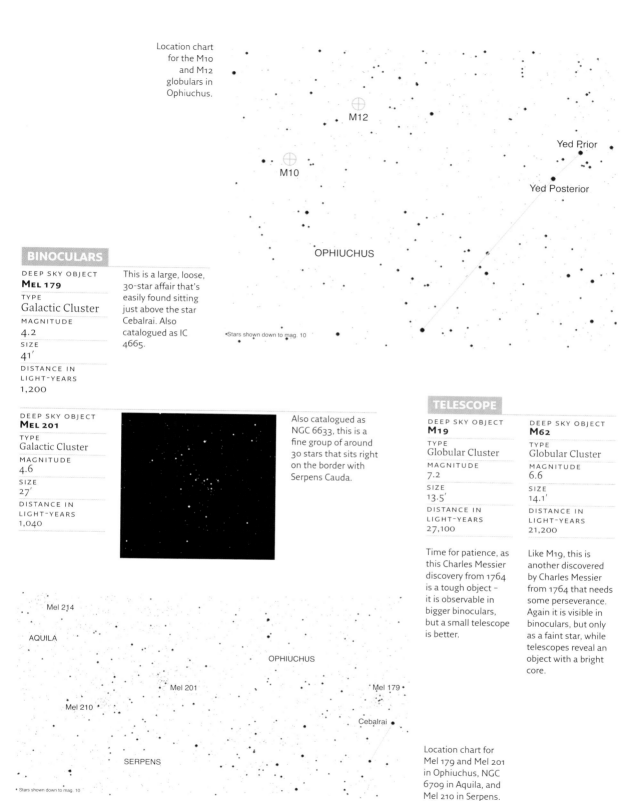

Location chart for the M10 and M12 globulars in Ophiuchus.

M12

M10

Yed Prior

Yed Posterior

OPHIUCHUS

*Stars shown down to mag. 10

DEEP SKY OBJECT
MEL 179

TYPE
Galactic Cluster

MAGNITUDE
4.2

SIZE
41'

DISTANCE IN
LIGHT-YEARS
1,200

This is a large, loose, 30-star affair that's easily found sitting just above the star Cebalrai. Also catalogued as IC 4665.

DEEP SKY OBJECT
MEL 201

TYPE
Galactic Cluster

MAGNITUDE
4.6

SIZE
27'

DISTANCE IN
LIGHT-YEARS
1,040

Also catalogued as NGC 6633, this is a fine group of around 30 stars that sits right on the border with Serpens Cauda.

DEEP SKY OBJECT
M19

TYPE
Globular Cluster

MAGNITUDE
7.2

SIZE
13.5'

DISTANCE IN
LIGHT-YEARS
27,100

Time for patience, as this Charles Messier discovery from 1764 is a tough object – it is observable in bigger binoculars, but a small telescope is better.

DEEP SKY OBJECT
M62

TYPE
Globular Cluster

MAGNITUDE
6.6

SIZE
14.1'

DISTANCE IN
LIGHT-YEARS
21,200

Like M19, this is another discovered by Charles Messier from 1764 that needs some perseverance. Again it is visible in binoculars, but only as a faint star, while telescopes reveal an object with a bright core.

Mel 214

AQUILA

OPHIUCHUS

Mel 201

Mel 210

Mel 179

Cebalrai

SERPENS

* Stars shown down to mag. 10

Location chart for Mel 179 and Mel 201 in Ophiuchus, NGC 6709 in Aquila, and Mel 210 in Serpens.

Serpens

LATIN NAME
Serpens
ENGLISH NAME
The Serpent
ABBREVIATION
Ser
LATIN
POSSESSIVE
Serpentis

α STAR
Unukalhai
MAGNITUDE
2.65
STAR COLOUR
Yellow

The Serpent is split into two groupings: Serpens Caput is its head (the western part), while Serpens Cauda is its tail (the eastern part). When searching for objects, make sure you're looking in the right bit, as some books only say 'Serpens', without specifying head or tail, and you could spend a while confused and getting cold – or is that just me?

The rest of the body, by the way, is wrapped around inside the constellation of Ophiuchus, the Serpent Bearer (see page 80).

Serpens Caput, the Serpent's Head

Serpens Cauda, the Serpent's Tail

BINOCULARS

DEEP SKY OBJECT
MEL 210
TYPE
Galactic Cluster
MAGNITUDE
4.6
SIZE
52'
DISTANCE IN
LIGHT-YEARS
1,300

Around 80 stars make up this cluster, with individual stars starting at eighth magnitude, then it's fainter all the way. Located in Serpens Cauda. Also catalogued as IC 4756. Look for it on the location chart in Ophiuchus on page 80.

The larger grouping of stars on the left is Mel 210, while on the right, and here for size comparison, is the smaller Mel 201 that's just over the border in Ophiuchus.

The width of the diagram is about five Moon diameters (~2.5 degrees), meaning that with most binoculars you should see them both in the same field.

DEEP SKY OBJECT
M5
TYPE
Globular Cluster
MAGNITUDE
5.6
SIZE
17'
DISTANCE IN
LIGHT-YEARS
24,500

A very old slightly oval-shaped globular that, under super skies, is apparently observable with the unaided eye – I haven't managed it, though. Located in Serpens Caput.

This cluster of about 100 stars in Serpens Cauda includes the great **EAGLE NEBULA**, which appears as a greyish smudge. Larger and larger telescopes, or those with imaging capabilities, will show a most irregular nebula and all-round wonderful sight – something like this image – hopefully.

DEEP SKY OBJECT
M16

TYPE
Galactic Cluster

MAGNITUDE
6.0

SIZE
7′

DISTANCE IN
LIGHT-YEARS
7,000

DOUBLE STAR
ALYA
θ Ser

MAGNITUDE
4.6 & 5.4

SEPARATION
22.1″

COLOURS
Both white

Location chart for the M5 globular in Serpens Caput.

Unukalhai

SERPENS

M5

Stars shown down to mag. 10

Scutum

LATIN NAME
Scutum
ENGLISH NAME
The Shield
ABBREVIATION
Sct
LATIN
POSSESSIVE
Scuti

α STAR
Sobieski
MAGNITUDE
4.0
STAR COLOUR
Yellow

The Shield is the fifth smallest constellation, and the only way of being noticed if you're that small is to be somewhere where you're bound to be noticed. In space terms that would have to be in front of the Milky Way – you then become a little gem. This is where Johann Hevelius built Scutum. He knew exactly what he was doing – the place is filled with star clouds galore.

BINOCULARS

DEEP SKY OBJECT
M11
TYPE
Galactic Cluster
MAGNITUDE
5.8
SIZE
14'
DISTANCE IN
LIGHT-YEARS
6,000

With a really dark sky the **WILD DUCK CLUSTER** can just about be glimpsed with the unaided eye. Binoculars show what is considered to be one of the finest clusters of the skies as a smudge, and a telescope begins to resolve some of the 3,000 or so stars that form this compact group. To me, the main irregular groupings of stars make the shape of a roast chicken – it's still game.

Corona Borealis

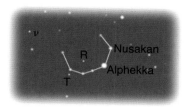

A nice, happy, curving group of seven stars that represent the crown given to Ariadne by Dionysus. He then threw it into the sky when she died. In another tale he threw the crown into the stars to prove to her that he was a god – the things some people do!

LATIN NAME
Corona Borealis
ENGLISH NAME
The Northern Crown
ABBREVIATION
CrB
LATIN POSSESSIVE
Coronae Borealis

α STAR
Alphekka
MAGNITUDE
2.23
STAR COLOUR
White

BINOCULARS

DOUBLE STAR
ν¹ AND ν² CrB
MAGNITUDE
5.2 & 5.4
SEPARATION
5′ 56″
COLOURS
Both reddish

After taking a look at the Great Cluster, M13 in Hercules, swing your binoculars south and you'll easily spot this fine double.

VARIABLE STAR
R CrB
Also TYC 2039-642-1 and HD 141352
MAGNITUDE RANGE
5.8 to 14.8
PERIOD
Irregular

Such a variable star that the magnitude range is itself variable! This is because we are looking at a 'sooty' star. The brightness falls as soot builds up in its atmosphere. When there's too much soot the star coughs and clears it away. You could say it likes to sweep away the sooty show.

VARIABLE STAR
T CrB
also TYC 2037-1144-1 and HD 143454
MAGNITUDE RANGE
10.8 to 2.0
PERIOD
About every 9,000 days!

Known as the **BLAZE STAR**, here is a recurrent nova-type variable that bursts into life every now and then. Past flarings have happened in 1866 and 1946, so the next could be any day now.

October to December Skies

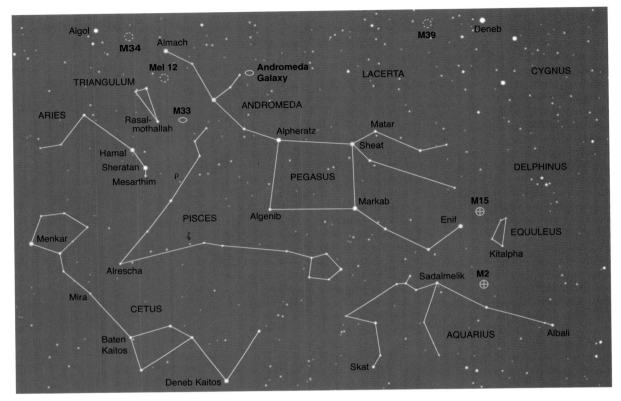

- Algol
- M34
- Almach
- Mel 12
- TRIANGULUM
- ARIES
- Rasal-mothallah
- M33
- Hamal
- Sheratan
- Mesarthim
- ρ
- Menkar
- Alrescha
- Mira
- CETUS
- Baten Kaitos
- Deneb Kaitos
- PISCES
- ζ
- Algenib
- Andromeda Galaxy
- ANDROMEDA
- Alpheratz
- PEGASUS
- Markab
- LACERTA
- Matar
- Sheat
- Enif
- Sadalmelik
- Skat
- AQUARIUS
- M39
- Deneb
- CYGNUS
- DELPHINUS
- M15
- EQUULEUS
- Kitalpha
- M2
- Albali

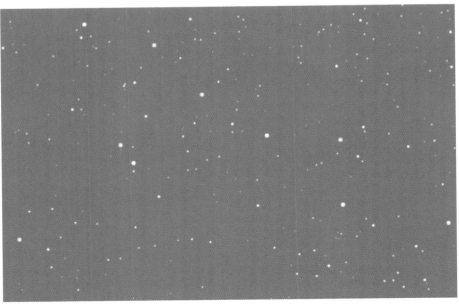

Galactic Cluster
Globular Cluster
Nebula
Planetary Nebula
Galaxy

The northern autumn skies looking south.

Cassiopeia

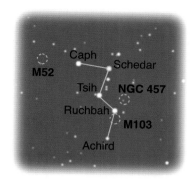

This outspoken Greek queen, the wife of Cepheus, the King (see overleaf), rolls around close to the pole of the sky, making her always visible for a large part of the northern hemisphere. The classic 'W' asterism of the main stars makes her easily identifiable, and, as she sits in the Milky Way, this is a great area to have a look at her in binoculars.

LATIN NAME
Cassiopeia
ENGLISH NAME
The Queen
ABBREVIATION
Cas
LATIN POSSESSIVE
Cassiopeiae

α STAR
Schedar
MAGNITUDE
2.2
STAR COLOUR
Yellow

EYE

This unstable star can change brightness quite quickly, so it's worth keeping an eye on it if you're out stargazing and Cassiopeia is visible. This is an example of an irregular-period variable star.

VARIABLE STAR
TSIH
γ Cas
MAGNITUDE RANGE
1.6 to 3.0
PERIOD
~0.7 days

DEEP SKY OBJECT
MEL 7
TYPE
Galactic Cluster
MAGNITUDE
6.4
SIZE
13'
DISTANCE IN LIGHT-YEARS
9,000

The stars of Mel 7 look like a skier to me (okay, the poles are a bit short), but an alternative name is the Owl Cluster – I guess that might be more appropriate for the night sky. Twit-twoo.

This is a great cluster with a variety of around 100 brighter and fainter stars forming an excellent shape. The brightest star in the field (one of the 'eyes'), φ Cas (Phi), is not actually a member of the group but rather around 6,700 light-years closer to us – this just proves that everything we see is so far away that it all appears to be at the same 'fixed' distance, when in reality the gaps between things can be vast. Also catalogued as NGC 457.

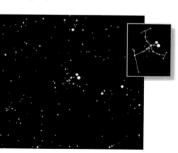

BINOCULARS

DEEP SKY OBJECT
M103
TYPE
Galactic Cluster
MAGNITUDE
7.4
SIZE
6'
DISTANCE IN LIGHT-YEARS
8,000

A smally but a goody with its fan-shaped appearance, M103 is composed of, extremely roughly, 60 stars, the brightest of which resolve from the fuzz in binoculars, but a telescope gives the best view.

DEEP SKY OBJECT
M52
TYPE
Galactic Cluster
MAGNITUDE
7.3
SIZE
13'
DISTANCE IN LIGHT-YEARS
5,000

Like M103, this 40-ish strong group also has a kind of fan-shaped appearance.

Cepheus

LATIN NAME
Cepheus
ENGLISH NAME
The King
ABBREVIATION
Cep
LATIN
POSSESSIVE
Cephei

α STAR
Alderamin
MAGNITUDE
2.44
STAR COLOUR
White

Cepheus was the son of Phoenix, King of Ethiopia. Everything seemed to be going well when he took over the crown, but then, due to the non-stop bragging of his wife, Cassiopeia, he was obliged to chain his daughter, Andromeda, to a rock as a sacrifice to Cetus, the sea monster. Thankfully, along came Perseus in his winged sandals to rescue her, but it was touch and go for a while.

EYE

VARIABLE STAR
δ Cep
MAGNITUDE RANGE
3.9 to 5.0
PERIOD
5.3663 days

A most interesting yellow star in that it is variable with an exact period. This star was the first of a whole new class of variables named Cepheids: stars whose magnitude fluctuates precisely with their period.

VARIABLE STAR
μ Cep
MAGNITUDE RANGE
3.43 to 5.1
PERIOD
~730 days

Named the **GARNET STAR** by William Herschel due to its intense deep-red colour, this is a semi-regular variable type.

DEEP SKY OBJECT
NGC 1499
TYPE
Nebula
MAGNITUDE
5.0
SIZE
2° 40' × 40'
DISTANCE IN
LIGHT-YEARS
1,000

The **CALIFORNIA NEBULA**. Just visible with the eye and, yes, with more power it is sort of shaped like that west coast USA state. This is located on the northern winter chart as well as on the North Celestial Pole chart.

Andromeda

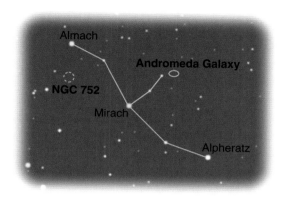

After that sea-monster escapade – being chained to a rock as a sacrifice after her mother, Cassiopeia, had said her daughter was more beautiful than the Nereids, which cheesed off their father Poseidon (Andromeda wasn't too chuffed either, mind you) – it was no surprise that Andromeda didn't want to go back home. She flew with Perseus to Argos to pick up a cheap fridge, some cutlery and a chariot.

LATIN NAME
Andromeda
ENGLISH NAME
Andromeda
ABBREVIATION
And
LATIN POSSESSIVE
Andromedae

α STAR
Apheratz
MAGNITUDE
2.06
STAR COLOUR
White

EYE

The **GREAT NEBULA** is the furthest realistic object you can see with your eye (see M33 in Triangulum on page 92). It may look like an insignificant fuzzy patch, but remember that what you are seeing is a galaxy larger than our own at a distance of over 2.8 million light-years (at current estimates). Of course you remember this distance is also 26 quintillion kilometres in Earth units! It's a mighty long way. Even so, it's enormous: 3 degrees in length – that's six times the width of the Moon.

DEEP SKY OBJECT
M31
TYPE
Galaxy
MAGNITUDE
4.8
SIZE
3°
DISTANCE IN LIGHT-YEARS
~2.8 million

M31 has two fuzzy friends: the one in front of its spiral arms on the lower left is the dwarf elliptical galaxy M32 and the one hanging around to the top right is the spheroidal galaxy M110.

BINOCULARS

DEEP SKY OBJECT
MEL 12
TYPE
Galactic Cluster
MAGNITUDE
5.5
SIZE
50'
DISTANCE IN LIGHT-YEARS
1,300

Around 70 stars form this loose group that appears attached (but isn't) to a curving chain of brighter coloured stars. It looks great in binoculars, and it's best found using Triangulum. Also known as NGC 752. See the location chart in Triangulum on page 92.

Hooray, the second greatest double star in the sky – Albireo (β Cyg) being the best.

TELESCOPE

DOUBLE STAR
ALMACH
γ And
MAGNITUDE
2.3 & 5.1
SEPARATION
10"
COLOURS
Yellow & green

Perseus

LATIN NAME
Perseus
ENGLISH NAME
Perseus
ABBREVIATION
Per
LATIN POSSESSIVE
Persei

α STAR
Mirfak
MAGNITUDE
1.8
STAR COLOUR
Yellow

Zeus, as you may have read, just could not help changing himself into some creature or other in order to seduce a lady. This time, however, his transformation was into...rain! Clearly the maiden Danae was impressed, because nine months later along came Perseus. Rain? Anyway, when Perseus was big, as a reward for cutting off the head of Medusa, Athena placed him in a splendid northern part of the Milky Way.

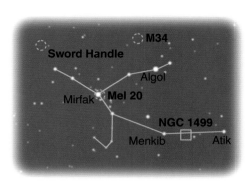

EYE

VARIABLE STAR
ALGOL
α Per
MAGNITUDE RANGE
2.1 to 3.4
PERIOD
2 days 20 hours 48 minutes

An eclipsing binary where you can watch the light levels change over the course of an evening – which is why this also has the name of the 'Winking Demon'. The minimum magnitude lasts for ten hours. This was the second binary star discovered in 1667 by Geminiano Montaniari.

DEEP SKY OBJECT
MEL 13 & 14
TYPE
Galactic Clusters
MAGNITUDE
4.3 & 4.4
SIZE
30' each
DISTANCE IN LIGHT-YEARS
7,100 & 7,400

The so-called **SWORD HANDLE** is a wondrous double galactic cluster of magnitudes 4.4 and 4.7 respectively. They are both half a degree in diameter and together are a great find. Although they are easily visible to the unaided eye, try sweeping the area with binoculars and it's Milky Way stars and patches all the way. In the picture, Mel 14 is the one on the left. Also designated as NGC 869 & 884.

DEEP SKY OBJECT
M34
TYPE
Galactic Cluster
MAGNITUDE
5.2
SIZE
35'
DISTANCE IN LIGHT-YEARS
1,400

Tricky indeed, with dark skies a must if you're to see this group of around 100 stars with the eye.

DEEP SKY OBJECT
MEL 20
TYPE
Galactic Cluster
MAGNITUDE
2.9
SIZE
3°
DISTANCE IN LIGHT-YEARS
600

Known excitingly as the α **PERSEUS MOVING CLUSTER**, this is a sprawling family of stars based around the star Algol, which makes it very easy to find.

Pegasus

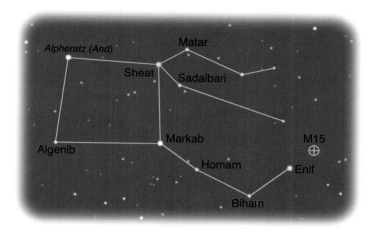

Alpheratz (And)
Matar
Sheat
Sadalbari
Markab
Algenib
Homam
Enif
M15
Biham

This is a story of unrequited love. Queen Antia, wife of Proteus, once looked fondly (if you know what I mean) upon one of their guests, Bellerophon. As he wasn't interested in the slightest, he left the kingdom on a flying horse, which he fell off (and if there's one thing you don't want to fall off, it's a flying horse). He met the ground at some speed while the horse made it up into the stars.

LATIN NAME
Pegasus
ENGLISH NAME
The Flying Horse
ABBREVIATION
Peg
LATIN POSSESSIVE
Pegasi

α STAR
Markab
MAGNITUDE
2.49
STAR COLOUR
White

BINOCULARS

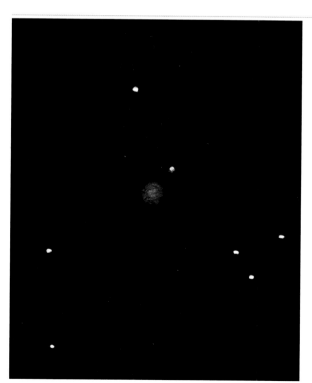

DEEP SKY OBJECT
M15
TYPE
Globular Cluster
MAGNITUDE
6.2
SIZE
12′
DISTANCE IN LIGHT-YEARS
33,600

Looking like a slightly more compact version of M13 in Hercules, this is an easy binocular object. It is believed to have undergone 'core collapse' – exciting I know – when the central stars lost a battle with gravity and now there's possibly a big black hole there. You'll need a good telescope to begin resolving the stars.

My drawing of M15 as observed on 18 September 1981.

Triangulum

LATIN NAME
Triangulum
ENGLISH NAME
The Triangle
ABBREVIATION
Tri
LATIN POSSESSIVE
Trianguli

α STAR
Rasalmothallah
MAGNITUDE
3.41
STAR COLOUR
White

The Greek name of the god Zeus begins with the letter delta (Δ) and this could well be the idea behind this constellation. Others, meanwhile, associate it with the shape of the island of Sicily or the outline of Egypt. It was certainly the case that the Egyptians reflected many earthly patterns in the heavens and vice versa – i.e. they were both inextricably linked – so that explanation is my favourite.

Rasal-mothallah

OBJECT
ANTON 2
TYPE
Old Constellation

Here is the abandoned constellation of **TRIANGULUM MINOR** which is made up of three fainter stars below the main constellation. Johann Hevelius popped this faint group in the sky and it appeared on many star atlases before the authorities said, 'Adios.'

DEEP SKY OBJECT
M33
TYPE
Galaxy
MAGNITUDE
5.7
SIZE
1°
DISTANCE IN LIGHT-YEARS
3 million

Witnesses claim to have seen M33 with the unaided eye on extremely clear nights. I have not, and I suspect them of drinking too much carrot juice. This is a good binocular object in crystal skies; you can make out an oval shape with a brighter centre.

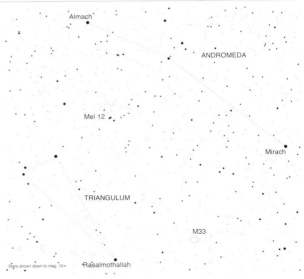

Location chart for the galaxy M33 in Triangulum and Mel 12 in Andromeda.

Aries

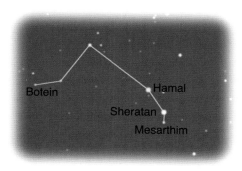

This ram was an immortal creature and, in ancient times, bad things happened to immortal things. Not only that, it had a golden fleece. Really, it was asking for trouble on several fronts. There are numerous stories of it flying and losing its fleece; one of the most memorable involves Jason and his Argonauts.

LATIN NAME
Aries
ENGLISH NAME
The Ram
ABBREVIATION
Ari
LATIN POSSESSIVE
Arietis

α STAR
Hamal
MAGNITUDE
2.0
STAR COLOUR
Yellow-orange

OBJECT
ANTON 3
TYPE
Old Constellation

Now flies annoy a lot of people, so maybe it was a good idea to use environmentally friendly, non-CFC, GM free, non-radioactive, nice smelling spray to remove this insect from the starry skies.

The abandoned constellation of **MUSCA BOREALIS**, the Northern Fly, can be found in the chart with ANTON 2 shown in Triangulum opposite.

DOUBLE STAR
MESARTHIM
γ Ari
MAGNITUDE
4.8 & 4.8
SEPARATION
8″
COLOURS
Both white

TELESCOPE

In 1664 this became one of the first double stars to be seen in a telescope. The English scientist Robert Hooke made the discovery while doing a bit of comet spotting.

Aquarius

LATIN NAME
Aquarius
ENGLISH NAME
The Water
Bearer
ABBREVIATION
Aqu
LATIN
POSSESSIVE
Aquarii

α STAR
Sadalmelik
MAGNITUDE
3.0
STAR COLOUR
Yellow

One story is that Aquarius could represent Ganymede, who was taken from his parents to be the cup-bearer of Zeus – the nectar of the gods flows from Aquarius' jug. Alternatively, he could be Deucalion, depicting the story of the great flood that happened during his time as king.

BINOCULARS

DEEP SKY OBJECT
M2
TYPE
Globular Cluster
MAGNITUDE
6.5
SIZE
13'
DISTANCE IN
LIGHT-YEARS
37,500

A fairly easy binocular object (I say fairly just in case you can't find it), looking like a small, featureless blurry star.

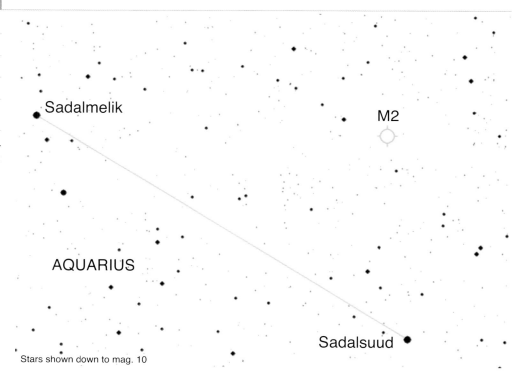

Location chart for the globular M2 in Aquarius.

Stars shown down to mag. 10

Again with the wonders of technology the colours of the Universe can be seen in all their glory.

DEEP SKY OBJECT
NGC 7293

TYPE
Planetary Nebula

MAGNITUDE
7.3

SIZE
16'

DISTANCE IN LIGHT-YEARS
450

The **HELIX NEBULA**, as it's called, is visually the biggest planetary nebula. Big because it's possibly the closest to us as well. A small telescope with non-light-polluted skies will show it as a fuzzy disc.

● Skat

AQUARIUS

Location chart for the Helix Nebula (NGC 7293) in Aquarius.

NGC 7293

Stars shown down to mag. 10

Pisces

LATIN NAME
Pisces
ENGLISH NAME
The Fish
ABBREVIATION
Psc
LATIN POSSESSIVE
Piscium

α STAR
Alrescha
MAGNITUDE
3·79
STAR COLOUR
White

Fish, fish, fishy fish. Two fish found a big egg in a river and when they brought it out onto the land (the story doesn't mention how the fish did land the thing) it hatched and a lovely dove flew out. What a smashing story.

EYE
DOUBLE STAR
ρ and **94**
MAGNITUDE
5.3 & 5.6
SEPARATION
7′ 27″
COLOURS
Yellowish & golden

TELESCOPE
DOUBLE STAR
ζ **Psc**
MAGNITUDE
5.6 & 6.5
SEPARATION
24″
COLOURS
White & yellowish

Cetus

The Egyptians saw this area of the sky as a crocodile, but the Greeks had to go one better and made a super-scary sea monster that rose out of the water in an attempt to devour Princess Andromeda. Thankfully she was saved by Perseus, who sorted out the monster once and for all.

LATIN NAME
Cetus
ENGLISH NAME
The Whale
ABBREVIATION
Cet
LATIN POSSESSIVE
Ceti

α STAR
Menkar
MAGNITUDE
2.54
STAR COLOUR
Red

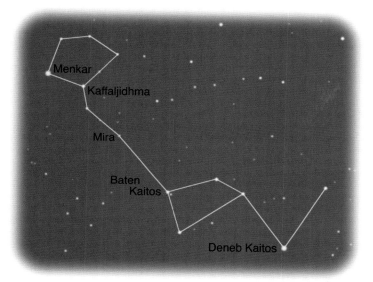

EYE & BINOCULARS

VARIABLE STAR
MIRA
o Cet

MAGNITUDE RANGE
2.0 to 10.1

PERIOD
331.96 days

Mira ('The Wonderful') was the first variable star identified – other than a nova – and so other long-period variables are also known as Mira-type stars. Dutch astronomer David Fabricus claimed the credit for this in 1596. As Mira fluctuates in size, its colour slowly changes too, from red to crimson.

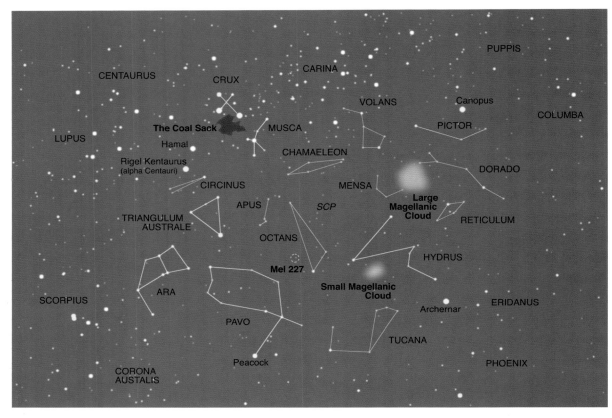

PUPPIS

CENTAURUS

CARINA

CRUX

VOLANS

Canopus

COLUMBA

The Coal Sack

MUSCA

PICTOR

LUPUS

Hamal

CHAMAELEON

DORADO

Rigel Kentaurus
(alpha Centauri)

CIRCINUS

MENSA

Large
Magellanic
Cloud

APUS

SCP

RETICULUM

TRIANGULUM
AUSTRALE

OCTANS

HYDRUS

Mel 227

Small Magellanic
Cloud

SCORPIUS

ARA

Archernar

ERIDANUS

PAVO

TUCANA

Peacock

PHOENIX

CORONA
AUSTALIS

For all the people
and animals
(including penguins)
living in or visiting
the southern
hemisphere.

The
Southern
Charts

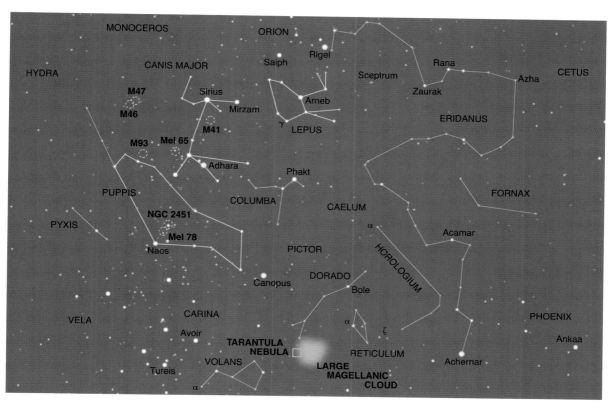

- Galactic Cluster
- ⊕ Globular Cluster
- ▢ Nebula
- ◇ Planetary Nebula
- ⬯ Galaxy

The southern
summer skies
looking south.

Canis Major

LATIN NAME
Canis Major
ENGLISH NAME
The Great Dog
ABBREVIATION
CMa
LATIN POSSESSIVE
Canis Majoris

α STAR
Sirius
MAGNITUDE
-1.46
STAR COLOUR
White

STAR
Sirius
DISTANCE
8.6 light-years
ABSOLUTE MAGNITUDE
1.4
LUMINOSITY
26
SPECTRAL CLASS
A1

STAR
Adhara (ε CMa)
DISTANCE
570 light-years
APPARENT MAGNITUDE
1.50
ABSOLUTE MAGNITUDE
-4.8
LUMINOSITY
8,000
SPECTRAL CLASS
B2

As well as the obvious hunting dog of Orion, other sagas tell of a poor canine that seems to have been passed from one owner to another, including Zeus, who employed it to look after Europa before sticking it in the sky. As befits its doggy nature, this is a well-behaved constellation containing Sirius (Greek for 'scorching'), the brightest star in the night-time sky.

EYE

DEEP SKY OBJECT
M41
TYPE
Galactic Cluster
MAGNITUDE
4.5
SIZE
38'
DISTANCE IN LIGHT-YEARS
2,300

Larger than the Moon, this is a happy family of around 100 bright and faint stars of different colours. It can be found on the location chart in Puppis on page 103.

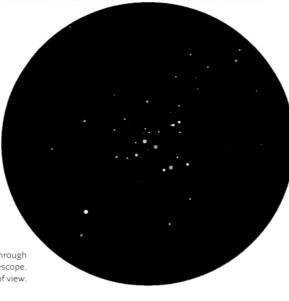

M41 through a telescope. 1¼° field of view.

DEEP SKY OBJECT
MEL 65
TYPE
Galactic Cluster
MAGNITUDE
4.1
SIZE
8'
DISTANCE IN LIGHT-YEARS
5,000

A truly fine cluster of around 60 stars make up this, the **TAU CANIS MAJORIS CLUSTER**, or call it NGC 2362 if you wish. See page 103 for the location chart.

Lepus

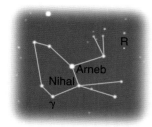

This hare is supposed to be a representation of the speedy god Hermes. Are they saying Hermes had big ears? Whatever, I like rabbits and hares – they never strike me as being dangerous animals. I had a rabbit once and I have nothing but good words for them.

LATIN NAME	Lepus
ENGLISH NAME	The Hare
ABBREVIATION	Lep
LATIN POSSESSIVE	Leporis

α STAR	
	Arneb
MAGNITUDE	2.6
STAR COLOUR	Yellowish

BINOCULARS

DOUBLE STAR	γ LEP
MAGNITUDE	3.7 & 6.3
SEPARATION	1′ 36″
COLOURS	Yellow & orange

VARIABLE STAR	
	R LEP
MAGNITUDE RANGE	5.5 to 11.7
PERIOD	~430 days

TELESCOPE

In 1845, English astronomer John Hind described this star as 'like a drop of blood on a black field' and so this very red point of light (more noticeable when at its faintest) became known as **HIND'S CRIMSON STAR**.

Reticulum

This is a fine little rhomboidal-shaped group found quite easily as it sits close to the Large Magellanic Cloud. The full original and glorious name for this, the sixth smallest constellation in the skies, was Reticulum Romboidalis, the Rhomboidal Net. This great-sounding title more than makes up for the fact that its brightest star is just magnitude 3.4, which means overall it's not a stunning constellation. Therefore, it's worth learning where it is just so you can say, 'There's Reticulum Romboidalis!' whenever you get the chance. If you can say it in a deep voice and roll your Rs at the start of the words, so much the better.

LATIN NAME	Reticulum
ENGLISH NAME	The Net
ABBREVIATION	Ret
LATIN POSSESSIVE	Reticuli

α STAR	
	α Ret
MAGNITUDE	3.4
STAR COLOUR	Yellow

EYE

DOUBLE STAR	ζ RET
MAGNITUDE	5.2 & 5.5
SEPARATION	5′ 10″
COLOURS	Both yellow

Here's something in Reticulum for the dedicated amateur: it's the magnitude 11.1 barred-spiral galaxy, NGC 1559. This is located about 50 million light-years away and it contains nearly 10,000 million Suns-worth of stuff. That sounds big but, in fact, NGC 1559 is actually about seven times smaller than our home Milky Way Galaxy.

Puppis

LATIN NAME
Puppis
ENGLISH NAME
The Stern
ABBREVIATION
Pup
LATIN
POSSESSIVE
Puppis

α STAR
Naos
MAGNITUDE
2.3
STAR COLOUR
Bluish

Originally this was one part of the great talking ship, *Argo Navis*, that was put in the sky as a whole constellation by Athena. Eventually broken up by the French astronomer Nicholas La Caille (who made a bit of a habit of doing this sort of thing – you could get away with it in the eighteenth century) into three separate constellations, Carina, Vela and Puppis. This is the most northerly chunk – can you see the splinters of its Argonautical encounter with 'the clashing rocks'?

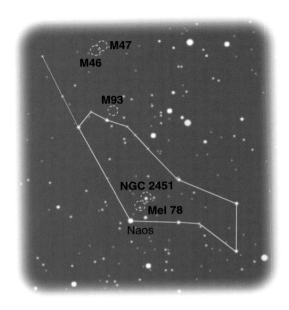

EYE

DEEP SKY OBJECT
M47
TYPE
Galactic Cluster
MAGNITUDE
4.4
SIZE
30'
DISTANCE IN
LIGHT-YEARS
1,600

This group of around 40 stars looks like a fuzzy smudge just smaller than the Moon in really dark skies. Binoculars will show its irregular nature, which you can compare using binoculars with the extremely faint roundish M46 next door.

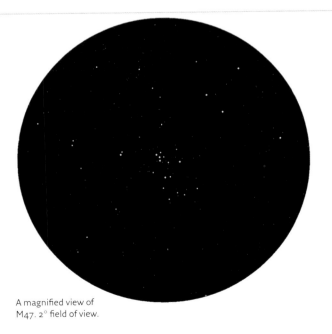

A magnified view of M47. 2° field of view.

DEEP SKY OBJECT
NGC 2451
TYPE
Galactic Cluster
MAGNITUDE
2.8
SIZE
50'
DISTANCE IN
LIGHT-YEARS
850

You can see this easily with the eye and yet, apparently, the great astronomers Charles Messier and William Herschel couldn't find it! Around 40 stars make up this fine group.

DEEP SKY OBJECT
M93

TYPE
Galactic Cluster

MAGNITUDE
6.2

SIZE
22'

DISTANCE IN
LIGHT-YEARS
3,600

A small but still bright little group attempting a kind of squashed Z-shape formation, or maybe an alien shuttle-craft. Whatever. The two brightest stars are eighth magnitude with a golden hue – they could be photon torpedoes fired from the craft that have been frozen in time. Yes, I think that's the most probable cause.

DEEP SKY OBJECT
M46

TYPE
Galactic Cluster

MAGNITUDE
6.1

SIZE
27'

DISTANCE IN
LIGHT-YEARS
5,400

To the eye, M46 looks just like a brighter patch of the Milky Way. It's a fair-sized cloud of similarly faintish stars. For those with a telescope the cluster is superbly enhanced by the addition of a tenth magnitude, 1-arc-minute-sized planetary nebula (NGC 2438) that's almost twice the distance of M46. (Image courtesy N.A.Sharp/NOAO/AURA/NSF)

DEEP SKY OBJECT
MEL 78

TYPE
Galactic Cluster

MAGNITUDE
5.7

SIZE
20'

DISTANCE IN
LIGHT-YEARS
4,200

A cluster of over 200 faintish stars in a roundish formation – so it looks like a weak globular cluster with chains and loops of stars within. Also catalogued as NGC 2477.

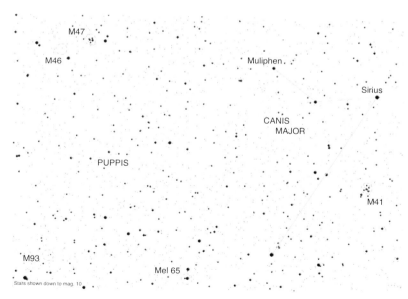

Location chart for M46, M47 and M93 in Puppis, with M41 and Mel 65 in Canis Major.

Dorado

LATIN NAME
Dorado
ENGLISH NAME
The Goldfish
ABBREVIATION
Dor
LATIN
POSSESSIVE
Doradus

α STAR
Bole
MAGNITUDE
3.3
STAR COLOUR
Bluish

This isn't a large constellation, but it's easy to find because sitting right across the southern border with Mensa, the Table, is the Large Magellanic Cloud. I've seen it called Xiphias, the Swordfish, in some older star charts, but that translation is completely wrong.

EYE

DEEP SKY OBJECT
LARGE MAGELLANIC CLOUD

TYPE
Barred Spiral Galaxy
MAGNITUDE
0.4
SIZE
$10° \times 9°$
DISTANCE IN LIGHT-YEARS
179,000

The Large Magellanic Clould is just a stunning piece of artwork, with an incredible size: over 20 times the diameter of the Moon! The bright small nebulous patch to the lower right of the Cloud is the Tarantula Nebula, visible to the eye as a fuzzy star.

One quarter the size of our Galaxy, this cloud, like the Small Magellanic, looks just like a large piece that has been taken away from the Milky Way and cast adrift. It's a satellite of our Galaxy and the fourth largest member in our Local Group (after us, M31 and M33). It's referred to by those in the know simply as the LMC. Binoculars will easily resolve the stars from the fuzz, and telescopes show a myriad of features.

DEEP SKY OBJECT
NGC 2070

TYPE
Emission Nebula

MAGNITUDE
5.0

SIZE
40′ × 20′

DISTANCE IN
LIGHT-YEARS
179,000

Lying within the Large Magellanic Cloud is the **TARANTULA NEBULA**, also known as 30 Doradus as it was first catalogued as a star. When you've seen all the nebulae in the world this one will be in your top five – guaranteed, or your money back. Amazingly its actual size is over 20 times that of the Orion Nebula! And still more amazingly – most nebulae we see are in our Galaxy and generally just a few thousand light-years away, while the Tarantula is in another galaxy at nearly 180,000 light-years away – *and* it's still visible to the unaided eye. In fact, it's the brightest emission nebula known.

A closer look at the Tarantula Nebula (above) as seen with someone else's 4m telescope – I seem to have lost mine. I'm sure I put it in the kitchen cupboard, but last time I looked it was gone. Anyway, it is thought that those long, streaming tarantula arms of hydrogen may stretch up to 1,800 light-years from the centre. This thing is absolutely whoppingly enormous and more. (Image courtesy NOAO/AURA/NSF)

April to June Skies

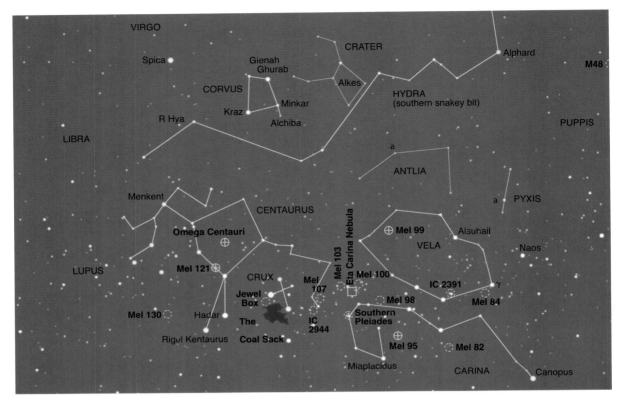

VIRGO

Spica

CRATER

Alphard

M48

Gienah
Ghurab

Alkes

CORVUS

HYDRA
(southern snakey bit)

PUPPIS

Minkar

Kraz

Alchiba

R Hya

LIBRA

a

ANTLIA

a PYXIS

Menkent

CENTAURUS

Alsuhail

LUPUS

Omega Centauri

⊕ Mel 99

VELA

Naos

Eta Carina Nebula

Mel 103

Mel 100

Mel 121 ⊕

IC 2391

γ

CRUX

Mel
107

Mel 98

Mel 84

Jewel
Box

Southern
Pleiades

Mel 130

Hadar

The

IC
2944

⊕
Mel 95

Mel 82

Rigel Kentaurus

Coal Sack

Miaplacidus

CARINA

Canopus

○ Galactic Cluster

⊕ Globular Cluster

□ Nebula

◇ Planetary Nebula

○ Galaxy

The southern
summer skies
looking south.

Hydra

LATIN NAME
Hydra
ENGLISH NAME
The Water Snake
ABBREVIATION
Hya
LATIN POSSESSIVE
Hydrae

α STAR
Alphard
MAGNITUDE
2.0
STAR COLOUR
Orange

This snaky monster was horrid on all levels – it had nine wriggling heads, poisonous breath and was so scary that lots of people died from looking at it. Hercules finally killed it, but it looked hopeless at first, because for every head he cut off three grew back!

This is the largest constellation in order of size, followed by Virgo and Ursa Major.

VARIABLE STAR
R HYA
Also TYC 6713-56-1 and HD 117205
MAGNITUDE RANGE
4.5 to 9.5
PERIOD
389 days

This is a Mira-type (long-period) variable.

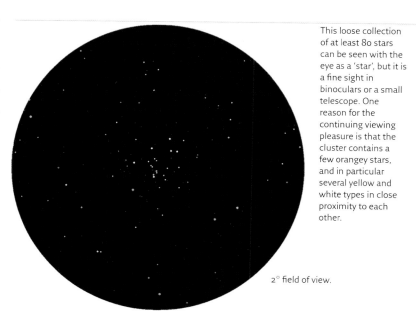

2° field of view.

This loose collection of at least 80 stars can be seen with the eye as a 'star', but it is a fine sight in binoculars or a small telescope. One reason for the continuing viewing pleasure is that the cluster contains a few orangey stars, and in particular several yellow and white types in close proximity to each other.

BINOCULARS

DEEP SKY OBJECT
M48
TYPE
Galactic Cluster
MAGNITUDE
5.8
SIZE
54'
DISTANCE IN LIGHT-YEARS
1,500

Centaurus

LATIN NAME
Centaurus

ENGLISH NAME
The Centaur

ABBREVIATION
Cen

LATIN POSSESSIVE
Centauri

α STAR
Rigel Kentaurus

MAGNITUDE
-0.01

STAR COLOUR
Yellow

STAR
Rigel Kentaurus

DISTANCE
4.3 light-years

APPARENT MAGNITUDE
-0.01 & +1.36

ABSOLUTE MAGNITUDE
4.4 & 5.7

LUMINOSITY
1.6 & 0.5

SPECTRAL CLASS
G2 & K1

STAR
Hadar (β Cen)

DISTANCE
320 light-years

APPARENT MAGNITUDE
0.6 (var.)

ABSOLUTE MAGNITUDE
-4.4

LUMINOSITY
10,000

SPECTRAL CLASS
B1

Half-man, half-horse, the Centaur Chiron taught Hercules until there was an 'accident' – some sort of party got out of hand, allegedly – and Hercules fired a poisoned arrow which struck the immortal Chiron. He would remain forever in pain unless Zeus could help. He did and the Centaur went up into the sky.

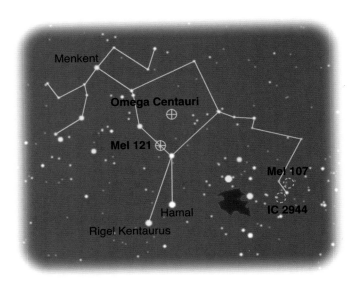

DEEP SKY OBJECT
MEL 118

TYPE
Globular Cluster

MAGNITUDE
3.65

SIZE
36′

DISTANCE IN LIGHT-YEARS
17,000

This is the 'star' **OMEGA CENTAURI**. How can a star be the best globular cluster in the sky? Well, it's because the true nature of this mysterious object wasn't known until the invention of the telescope – it simply looks like a star (a slightly fuzzy one, mind you). Now we know it's not just an ordinary globular either, but the most massive and luminous one of all those in our Galaxy. Also known as NGC 5139.

Slightly oval by nature, Omega Centauri's outer stars begin resolving in binoculars, and it's a wondrous sight in larger telescopes.

DEEP SKY OBJECT
MEL 107

TYPE
Galactic Cluster

MAGNITUDE
5.3

SIZE
12′

DISTANCE IN LIGHT-YEARS
5,500

This is a close, tightly formed family of about 100 stars which will begin to resolve in binoculars, though a telescope is really needed to see its full attractiveness. Also catalogued as NGC 3766.

DEEP SKY OBJECT
MEL 121

TYPE
Globular Cluster

MAGNITUDE
6.0

SIZE
14'

DISTANCE IN
LIGHT-YEARS
3,900

This blob has been described as 'magnificent' by those observing it in larger telescopes. There's a fourth-magnitude star in the field which just adds to the great scene. Also goes by NGC 5286.

There seems to be confusion in some quarters over what IC 2944 is. According to the NGC/IC database it is a group of stars, the main four of which make a nice line just below Lambda Centauri (λ Cen). However, the people at Hubble use it for the surrounding so-called Running Chicken Nebula. According to my sources this nebula should be IC 2948, but am I right? Who knows? Who cares? Have I started something? I hope so.

DEEP SKY OBJECT
IC 2944

TYPE
Galactic Cluster

MAGNITUDE
5.5

SIZE
25'

DISTANCE IN
LIGHT-YEARS
5,800

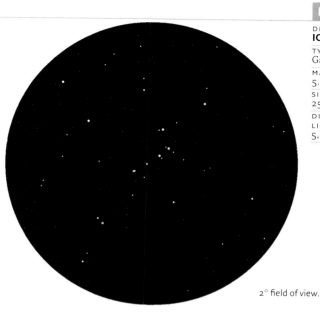

2° field of view.

Rigel Kentaurus

Hamal

Proxima Centauri

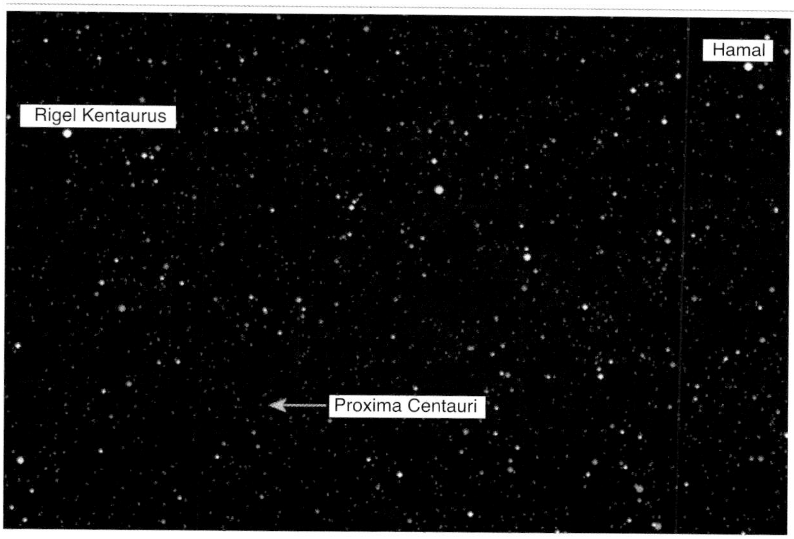

Proxima also has the designations α Centauri C, TYC 9010-4949-1 and HIP 70890.

Nearish to Rigel Kentaurus (α Cen) is the very small and faint Proxima Centauri: the nearest star to us after the Sun. It may be the third component of a system with the binary stars α Cen A and B at the 'centre'. So Proxima orbits these at a distance of some 13,000 AU, taking around 500,000 years to orbit once. Alternatively this may be a load of nonsense, and it's not part of the system at all. The location chart shows stars down to magnitude 11.5.

OBJECT
PROXIMA CENTAURI

TYPE
Red dwarf star (M5)

APPARENT MAGNITUDE
11.0

ABSOLUTE MAGNITUDE
15.5

DISTANCE IN
LIGHT-YEARS
4.27

Crux

LATIN NAME
Crux
ENGLISH NAME
The Cross
ABBREVIATION
Cru
LATIN
POSSESSIVE
Crucis

α STAR
Acrux
MAGNITUDE
0.8
STAR COLOUR
Bluish

STAR
Acrux (actually two stars)
DISTANCE
321 light-years
APPARENT
MAGNITUDE
1.4 + 2.1 (combined to 0.8)
ABSOLUTE
MAGNITUDE
-4.2
LUMINOSITY
2,500
SPECTRAL CLASS
B0.5 + B1

STAR
Mimosa (β Cru)
DISTANCE
353 light-years
APPARENT
MAGNITUDE
1.25 (var.)
ABSOLUTE
MAGNITUDE
-3.9
LUMINOSITY
34,000
SPECTRAL CLASS
B0

STAR
Gacrux (γ Cru)
DISTANCE
88 light-years
APPARENT
MAGNITUDE
1.6 (var.)
ABSOLUTE
MAGNITUDE
-0.56
LUMINOSITY
1,500
SPECTRAL CLASS
M4

This is the smallest constellation and yet it has an incredible variety of objects to look at due to a fine section of the Milky Way running behind the scene. The four bright stars are easily found and so have been used by seafarers through the ages as a southern equivalent of the Plough.

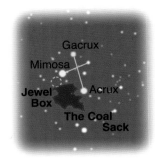

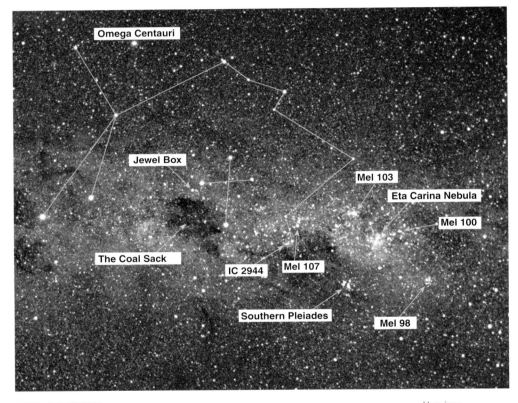

Here is an exceptional part of the entire night sky. Unaided-eye objects fill the wonderful Milky Way scene, with the two bright stars of Alpha and Beta Centauri sitting on the left. (Image courtesy of Peter Michaud)

DEEP SKY OBJECT
MEL 114
TYPE
Galactic Cluster
MAGNITUDE
4.2
SIZE
10'
DISTANCE IN
LIGHT-YEARS
7,600

The JEWEL BOX
Cluster is a 'gem' of a group. In binoculars, its bright multi-coloured stars glisten like a box of jewels, with blues, reds and whites galore. This cluster, also catalogued as NGC 4755, sits around Kappa Crucis (κ Cru) and only has to pop next door to the Coal Sack dark nebula if it ever wants a cup of tea. (Image courtesy NOAO/AURA/NSF)

DEEP SKY OBJECT
THE COAL SACK
TYPE
Dark Nebula
SIZE
6° 30' × 5°
DISTANCE IN
LIGHT-YEARS
550

This 'hole-looking thing' in the rich, flowing Milky Way is actually caused by a cloud of dust and gas that is blocking out the starlight from behind. The Sack is quite large and parts of it cross the borders into the constellations Centaurus and Musca.

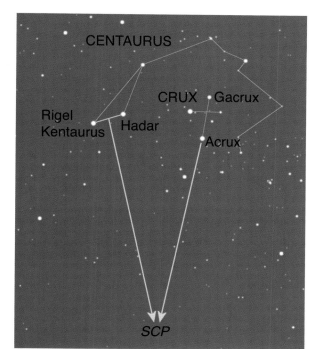

DOUBLE STAR
ACRUX
α Cru
MAGNITUDE
1.4 & 4.9
SEPARATION
1' 30"
COLOURS
Both bluish

DOUBLE STAR
GACRUX
γ Cru
MAGNITUDE
1.6 & 6.5
SEPARATION
1' 50"
COLOURS
Red & white

Here's how to use the Crux 'Pointers' together with Rigel Kentaurus and Hadar to find the South Celestial Pole (SCP) of the sky – alas there is no South Star waiting for you there.

Carina

LATIN NAME
Carina
ENGLISH NAME
The Keel
ABBREVIATION
Car
LATIN POSSESSIVE
Carinae

α Star
Canopus
MAGNITUDE
-0.72
STAR COLOUR
Yellowish

STAR
Canopus
DISTANCE
313 light-years
ABSOLUTE MAGNITUDE
-2.5
LUMINOSITY
200,000
SPECTRAL CLASS
A9

Until 1763, this keel, together with a stern (Puppis) and some sails (Vela), formed the great vessel *Argo Navis*, famed for crossing waters where no ship had gone before, with its crew Jason and the Argonauts. However, Nicholas 'Blackbeard' La Caille had sabotage on his mind and so we've ended up with separate constellations.

EYE

DEEP SKY OBJECT
MEL 82
TYPE
Galactic Cluster
MAGNITUDE
3.8
SIZE
30′
DISTANCE IN LIGHT-YEARS
1,300

Also designated NGC 2516, this is a bright cluster of about 100 stars, with spirally arms of stars that look somewhat as its common name suggests – **THE GARDEN SPRINKLER**. There are a couple of golden-coloured stars, which always make a good contrast in any cluster.

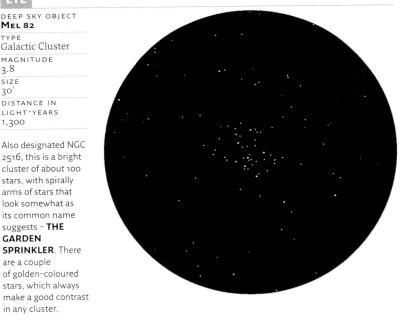

DEEP SKY OBJECT
MEL 98
TYPE
Galactic Cluster
MAGNITUDE
4.2
SIZE
35′
DISTANCE IN LIGHT-YEARS
3,000

Apparently there are 171 stars in this cluster. How anyone can be that precise is a mystery. Anyway, it is a great group in binoculars and with a telescope set to low power, due partly to several curving arch appearances of the stars. Also designated as NGC 3114.

DEEP SKY OBJECT
MEL 102

TYPE
Galactic Cluster

MAGNITUDE
1.9

SIZE
50'

DISTANCE IN
LIGHT-YEARS
479

This fine cluster around Theta Carinae, also designated as IC 2602, has affectionately been called the Southern Pleiades.

A close-up of the Eta Carina area shows how rich it is. The lower loose group of stars in the darker Milky Way is Mel 102, while the big cluster to the left of the nebula is Mel 103.

DEEP SKY OBJECT
MEL 103

TYPE
Galactic Cluster

MAGNITUDE
3.0

SIZE
55'

DISTANCE IN
LIGHT-YEARS
1,300

Also catalogued as NGC 3532, this starry group is near the Eta Carina Nebula (NGC 3372). It sits in a very busy part of the Milky Way, so add binoculars and the scene is marvellous.

DEEP SKY OBJECT
NGC 3372

TYPE
Nebula

MAGNITUDE
5.0

SIZE
2°

DISTANCE IN
LIGHT-YEARS
10,000

The Eta Carina Nebula is one of the great star-forming regions in our Galaxy. Here one of the most massive stars we have ever detected was created – Eta Carina – hence the name. The star is 4 million times brighter than the Sun with over 150 times the mass – not something that makes it all that stable. Bright wedges of nebulosity cut by dark lanes make this a fascinating telescopic object.

Only for fun, just take a closer look into the central bright heart of the Eta Carina Nebula – buried in here is the star Eta Carina itself, which until 1841 was the second brightest star in the sky. A violent eruption of gas put paid to that and the star disappeared underneath this exploded gas. This episode produced a new nebula (the round glob thing seen in the centre of the image) known as the Homunculus Nebula. (Image courtesy NOAO/AURA/NSF)

DEEP SKY OBJECT
MEL 100

TYPE
Galactic Cluster

MAGNITUDE
4.7

SIZE
6'

DISTANCE IN
LIGHT-YEARS
8,400

This young and small cluster of about 50 red and blue stars appears as a star-like patch to the eye, but splits wonderfully in binoculars. Also designated as NGC 3293, this is viewed as one of the finest star clusters, taking appearance and position into account – it sits just to the northwest of the Eta Carina Nebula.

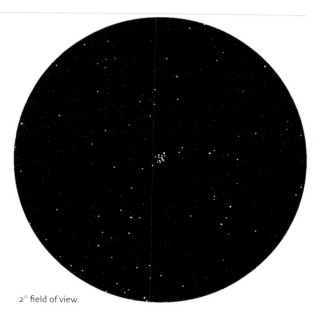

2° field of view.

DEEP SKY OBJECT
MEL 95

TYPE
Globular Cluster

MAGNITUDE
6.3

SIZE
13.8'

DISTANCE IN
LIGHT-YEARS
31,200

This globular can just be seen with the eye if you have good skies. No stars can actually be resolved in binoculars, but it does appear as a bright round fuzz. NGC 2808, as it is also known, is often described by those with a telescope as one the finest globulars around.

Vela

LATIN NAME
Vela
ENGLISH NAME
The Sails
ABBREVIATION
Vel
LATIN
POSSESSIVE
Velorum

α STAR
Suhail al Muhlif
MAGNITUDE
1.8
STAR COLOUR
Bluish

Ahoy! Here's another piece of the fine ship *Argo Navis*. In ye olden dayes the ship was complete, but now three pieces float free (see Puppis (page 102) and Carina (page 112)). Many writers include the constellation of Pyxis, the Compass, in the shipwreck caused by astro-mapping Nicholas La Caille, but it was never a part of the original ship. Don't believe everything you read in books.

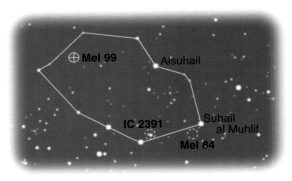

EYE

DEEP SKY OBJECT
IC 2391
TYPE
Galactic Cluster
MAGNITUDE
2.5
SIZE
50'
DISTANCE IN
LIGHT-YEARS
580

This large loose family of about 30 stars is centred around o **VEL**. It's easily located sitting just above the star Delta Velorum.

DEEP SKY OBJECT
MEL 84
TYPE
Galactic Cluster
MAGNITUDE
4.7
SIZE
20'
DISTANCE IN
LIGHT-YEARS
1,950

A fine group of around 50 stars discovered by La Caille and also catalogued as NGC 2547. Yes, he who 'did the carving up the ship deed'. It has a great vertical meander of stars through the cluster's centre, visible in binoculars.

2° field of view.

BINOCULARS

DEEP SKY OBJECT
MEL 99
TYPE
Globular Cluster
MAGNITUDE
6.8
SIZE
18'
DISTANCE IN
LIGHT-YEARS
16,300

This globular makes itself known in binoculars as a small fuzzy star. A telescope is needed to start resolving the stars. Also catalogued as NGC 3201.

DOUBLE STAR
SUHAIL AL MUHLIF
γ_2 Vel and γ_1 Vel
MAGNITUDE
1.9 & 4.2
SEPARATION
41"
COLOURS
Both bluish

July to September Skies

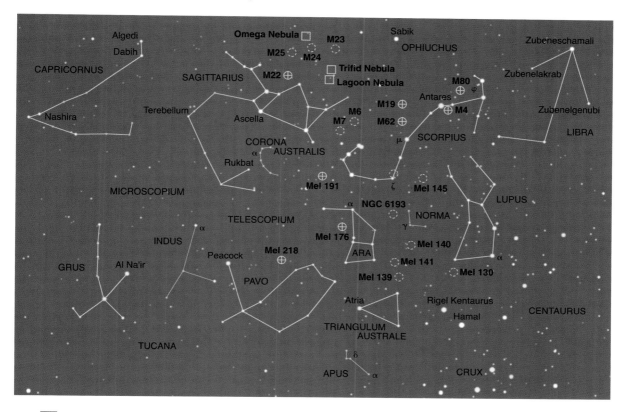

- ◌ Galactic Cluster
- ⊕ Globular Cluster
- ☐ Nebula
- ◇ Planetary Nebula
- ◯ Galaxy

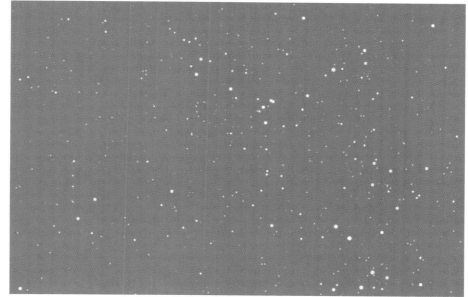

The southern winter skies looking south.

Capricornus

LATIN NAME
Capricornus
ENGLISH NAME
The Sea Goat
ABBREVIATION
Cap
LATIN
POSSESSIVE
Capricornus

α STAR
Algedi
MAGNITUDE
3.6
STAR COLOUR
Yellow

Capricornus, a curious mix of a goat with a fishy tail, is probably one of the oldest constellations of the zodiac. It is also the smallest, but this could be due to the fact that it used to be joined with Aquarius, forming the larger constellation known as the Ibex.

Here's the happy old watery-goat himself, as shown in the *Uranographia* star atlas of Johann Hevelius.

EYE

DOUBLE STAR
ALGEDI
α Cap
MAGNITUDE
4.2 & 3.6
SEPARATION
6′ 18″
COLOURS
Both goldishy (as opposed to goldfishy)

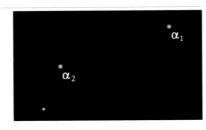

BINOCULARS

DOUBLE STAR
DABIH
β Cap
MAGNITUDE
3.1 & 6.1
SEPARATION
3′ 25″
COLOURS
Yellow & white

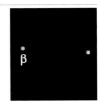

Libra

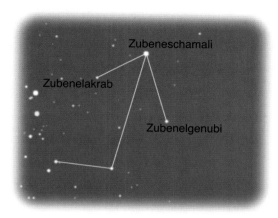

Libra is made from what used to be the claws of Scorpius, the Scorpion. They did this to make the 'scales of justice' as a way of thanking Julius Caesar for being such a decent chap – or so one story goes. In my opinion, Libra contains the best star name in the sky: Zubenelgenubi.

LATIN NAME
Libra
ENGLISH NAME
The Scales
ABBREVIATION
Lib
LATIN POSSESSIVE
Librae

α STAR
Zubenelgenubi
MAGNITUDE
2.75
STAR COLOUR
White

It's me as Isaac Newton (whilst filming for the BBC children's programme *Blue Peter*), just hanging by the front door of my/his house in Woolthorpe, Lincolnshire, England. Dashing,

I know. You may be wondering how this fits in here? Well, of course I'm waiting for it to get dark so I can use my new reflecting telescope to take a look at the fine double star Zubenelgenubi in Libra.

EYE

DOUBLE STAR
ZUBENELGENUBI
α_1 Lib and α_2 Lib
MAGNITUDE
2.8 & 5.2
SEPARATION
3' 51"
COLOURS
Bluish & white

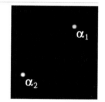

This great star name translates from Arabic as 'the southern claw of the scorpion'.

VARIABLE STAR
ZUBENELAKRIBI
δ Lib
MAGNITUDE RANGE
4.9 to 5.9
PERIOD
2.327 days

An Algol-type eclipsing variable. Of course, because of what Libra used to be, it isn't surprising that every named star is some part or other of a Scorpion's claw.

Scorpius

LATIN NAME
Scorpius
ENGLISH NAME
The Scorpion
ABBREVIATION
Sco
LATIN
POSSESSIVE
Scorpii

α STAR
Antares
MAGNITUDE
0.96
STAR COLOUR
Red

STAR
Antares
DISTANCE
604 light-years
ABSOLUTE
MAGNITUDE
-5.3
LUMINOSITY
7,500
SPECTRAL CLASS
M1

STAR
Shaula (λ Sco)
DISTANCE
~700 light-years
APPARENT
MAGNITUDE
1.63 (var.)
ABSOLUTE
MAGNITUDE
-3.5
LUMINOSITY
35,000
SPECTRAL CLASS
B1

In spite of Scorpius losing its claws to neighbouring Libra, this is still the best-looking constellation, thanks to the sweeping S-shaped curve of the stars that form the scorpion's body and round to the sting in its tail.

EYE

DEEP SKY OBJECT
M6
TYPE
Galactic Cluster
MAGNITUDE
4.2
SIZE
20′
DISTANCE IN
LIGHT-YEARS
1,600

3° field of view.

Around 80 stars here form such a pattern that M6 has become known as the **BUTTERFLY CLUSTER**. See what you think. It flaps just north of M7 against a darker area of the Milky Way, making it easy to see. The brightest of the group is the orangey-tinged semi-regular variable star BM Scorpii (mag. 5.5 to 7.0 over 850 days), which stands out nicely against all the other white-blue stars.

DEEP SKY OBJECT
M7
TYPE
Galactic Cluster
MAGNITUDE
3.3
SIZE
1° 20′
DISTANCE IN
LIGHT-YEARS
800

Sometimes known as **PTOLEMY'S CLUSTER** after he who described it back in AD 130. This is probably the finest deep-sky object in the constellation, and look at its size – more than twice as large as the Moon! To the eye the 80 bright luminaries that make up the cluster appear as a bright clump of Milky Way, but with optical aid they are set against the more distant Milky Way wash, making it simply a brilliant sight. (Image courtesy N.A.Sharp, REU rogram/NOAO/AURA/NSF)

DEEP SKY OBJECT
MEL 153

TYPE
Galactic Cluster

MAGNITUDE
2.5

SIZE
15′

DISTANCE IN
LIGHT-YEARS
5,900

Known as the **TABLE OF SCORPIUS**, this is an extremely young cluster of around 100 stars, perhaps only just over 3 million years old. Sitting close to Zeta Scorpii, the Table is a tightly knit group to the eye, and marvellous in binoculars or a small telescope. It is also catalogued as NGC 6231.

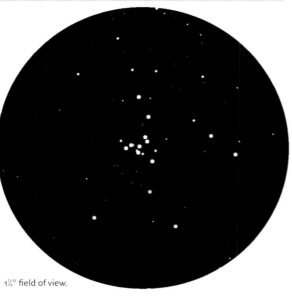

1¼° field of view.

DOUBLE STAR
ζ¹ AND ζ² Sco

MAGNITUDE
3.6 & 4.9

SEPARATION
6′ 30″

COLOURS
Orange & bluish

These two form the lower part of the Table of Scorpius. It's all a great part of the sky.

DOUBLE STAR
ω¹ and ω² Sco

MAGNITUDE
4.1 & 4.6

SEPARATION
14′ 30″

COLOURS
Blue & orange

DOUBLE STAR
μ Sco

MAGNITUDE
3.0 & 3.6

SEPARATION
5′ 30″

COLOURS
Both bluish

DEEP SKY OBJECT
M4

TYPE
Globular Cluster

MAGNITUDE
6.0

SIZE
14′

DISTANCE IN
LIGHT-YEARS
6,800

Not as compact as some globulars we know and love, but certainly one of the closest, and easy to find, as it sits just to the west of Antares. Try M4 in a telescope, it's globularlytastic.

DEEP SKY OBJECT
M80

TYPE
Globular Cluster

MAGNITUDE
7.3

SIZE
8.9′

DISTANCE IN
LIGHT-YEARS
32,600

Don't give up if you find this difficult to locate – so did Charles Messier, who discovered it in 1781. It's because M80 is bright but small, so it can be mistaken for a star. In 1999 the Hubble Space Telescope found this to be one of the most closely packed full-of-stars globulars that surround our Galaxy.

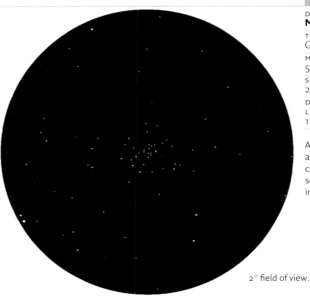

2° field of view.

DEEP SKY OBJECT
MEL 145

TYPE
Galactic Cluster

MAGNITUDE
5.8

SIZE
29′

DISTANCE IN
LIGHT-YEARS
1,650

Also catalogued as NGC 6124, this cluster of over 100 or so stars is quite loose in its appearance.

Sagittarius

LATIN NAME
Sagittarius
ENGLISH NAME
The Archer
ABBREVIATION
Sgr
LATIN
POSSESSIVE
Sagitarii

α STAR
Rukbat
MAGNITUDE
3.97
STAR COLOUR
Bluish-white

This is just a tremendous constellation – we're looking towards the centre of our Galaxy here, meaning the Milky Way is at its busiest. There is just so much to see with the eye and optical aids, and then there's the fact that the main stars look like a teapot! Inspired design.

EYE

OBJECT
ANTON 8
TYPE
Old Constellation

Straining to be seen above the Sagittarian Teapot is the abandoned constellation of Teabagus – the Great Teabag to the right and the Small Teabag to the left. The wondrous Milky Way here gives the two bags an almost perforated look.

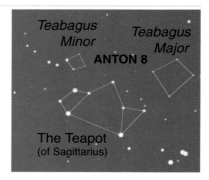

The fabulous Milky Way through Sagittarius.

DEEP SKY OBJECT
M8
TYPE
Emission Nebula
MAGNITUDE
5.8
SIZE
1° 30′ × 40′
DISTANCE IN
LIGHT-YEARS
5,200

This is the famous **LAGOON NEBULA**, which is just visible to your eye on the clearest of crispy nights. It gets its name from a dark band of lagoonish dust that meanders across the cloud and is visible in telescopes. Within the nebula is the young galactic cluster NGC 6530, which formed from the gas of the surrounding nebula.

DEEP SKY OBJECT
M22

TYPE
Globular Cluster

MAGNITUDE
5.1

SIZE
24'

DISTANCE IN
LIGHT-YEARS
10,000

An impressive globular – only beaten by one in Tucana (Mel 1) and one in Centaurus (NGC 5139). It's also great in binoculars, and impressive indeed in a telescope. M22 was probably the first object identified as a globular cluster.

DEEP SKY OBJECT
M24

TYPE
Star Clouds

MAGNITUDE
4.5

SIZE
1° 30'

DISTANCE IN
LIGHT-YEARS
10,000

The **SAGITTARIUS STAR CLOUD** is a patch of fuzzy light that is just a little brighter than the general mish-mash that is the Milky Way band. At three times the diameter of the Moon, it's a whopping good size too.

DOUBLE STAR
ARKAB PRIOR AND ARKAB POSTERIOR
β¹ and β² Sgr

MAGNITUDE
4.0 & 4.3

SEPARATION
28.3'

COLOURS
Bluish & white

These stars are so far apart that any diagram is too big to fit on the page! Being such an extremely wide double they are clearly identified on the main chart.

DEEP SKY OBJECT
M17

TYPE
Nebula

MAGNITUDE
7.0

SIZE
11'

DISTANCE IN
LIGHT-YEARS
5,000

This is the marvellous **OMEGA NEBULA**, named for its similarity to that Greek letter (ω) – it's also called the

HORSESHOE NEBULA for a further similarity reason. Meanwhile, in binoculars it looks like a piece of grey cake!

DEEP SKY OBJECT
M23

TYPE
Galactic Cluster

MAGNITUDE
5.5

SIZE
27'

DISTANCE IN
LIGHT-YEARS
2,150

This is a loose, slightly squashed-looking collection of faint stars, some of which are arranged in pleasant curvy-wurvy lines.

DEEP SKY OBJECT
M20

TYPE
Nebula

MAGNITUDE
9.0

SIZE
28'

DISTANCE IN
LIGHT-YEARS
5,200

Known as the **TRIFFID NEBULA** due to its three-part appearance, this can just be seen with binoculars as a white-green haze from non-light-polluted dark locations, but a telescope is best.

With not too difficult (!) CCD imaging through a telescope, the fine structures of M8, the Lagoon (below), and M20, the Triffid, can be revealed.

DEEP SKY OBJECT
M25

TYPE
Galactic Cluster

MAGNITUDE
4.6

SIZE
32'

DISTANCE IN
LIGHT-YEARS
2,000

Although it is visible to the unaided eye, you really need at least binoculars to do justice to this great cluster of around 30 stars.

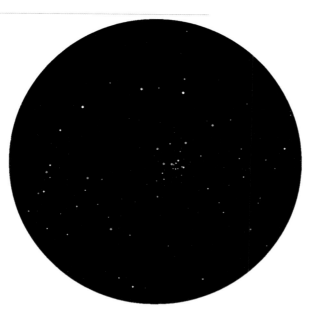

The two small curving rows of stars show the centre of M25. Then there's another larger curve of stars around the top left of the cluster.

2° field of view.

Corona Australis

LATIN NAME
Corona Australis
ENGLISH NAME
The Southern
Crown
ABBREVIATION
CrA
LATIN
POSSESSIVE
Coronae
Australis

α STAR
Alfecca
Meridiana
MAGNITUDE
4.1
STAR COLOUR
White

A faintish group that stands out not because of its brightness by any means, but because of its fine curvy shape. This probably just represents a crown that slipped off Sagittarius' head and now sits on the ground by his feet. I have an inkling he didn't like hats much and it was more of an 'accidental' slippage – you know what I mean?

BINOCULARS

DEEP SKY OBJECT
MEL 191
TYPE
Globular Cluster
MAGNITUDE
6.6
SIZE
13'
DISTANCE IN
LIGHT-YEARS
26,700

A good small object with brightness increasing towards a compact centre. It's also called NGC 6541. (Image courtesy NASA/STScI)

Triangulum Australe

Mel 139

Atria

The Dutch navigators Frederick de Houtman and Pieter Dirkszoon Keyser put this shape firmly on the starry maps in the late sixteenth century, but it could be much older as it is quite a distinctive group – much more so than the northern Triangulum.

LATIN NAME
Triangulum
Australe
ENGLISH NAME
The Southern
Triangle
ABBREVIATION
TrA
LATIN
POSSESSIVE
Trianguli
Australis

α STAR
Atria
MAGNITUDE
1.9
STAR COLOUR
Orange

BINOCULARS

DEEP SKY OBJECT
MEL 139
TYPE
Galactic Cluster
MAGNITUDE
5.1
SIZE
12′
DISTANCE IN
LIGHT-YEARS
2,700

Great, great and more great. A 60-star-strong group in a squashed kind of S shape that may just be glimpsed with the unaided eye. A part of this cluster crosses the border into the neighbouring constellation of Norma. Also known as NGC 6025. Trundle over to see the location chart in Norma overleaf for this object.

2° field of view.

Norma

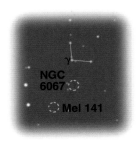

LATIN NAME
Norma
ENGLISH NAME
The Level
ABBREVIATION
Nor
LATIN
POSSESSIVE
Normae

α STAR
Nor
MAGNITUDE
3.0
STAR COLOUR
Bluish

Originally called Norma et Regula, the Level and Square, and attached to Lupus and Ara. But Nicholas 'carving up the constellations is my middle name' La Caille did his business in 1750 (see Carina, Vela and Puppis) and littered the sky with faint ghostly groups, of which Norma is one. However, we are in the heart of Milky Way Land, which makes up for everything.

BINOCULARS

DEEP SKY OBJECT
MEL 140
TYPE
Galactic Cluster
MAGNITUDE
5.6
SIZE
13'
DISTANCE IN
LIGHT-YEARS
4,600

A fine cluster of over 100 stars close to kappa Normae. It's an interesting object in a larger telescope, as some stars appear in curving chains with a hole in its centre – not a black hole, although it is black! Also catalogued as NGC 6067.

DEEP SKY OBJECT
MEL 141
TYPE
Galactic Cluster
MAGNITUDE
5.4
SIZE
12'
DISTANCE IN
LIGHT-YEARS
2,900

A great, uneven-looking, bright cluster, also catalogued as NGC 6087.

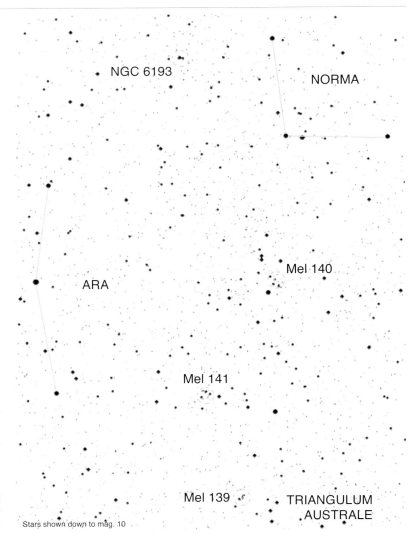

Location chart for the galactic clusters Mel 139 in Triangulum Australe, Mel 140 and Mel 141 in Norma, and NGC 6193 in Ara. Stars shown down to mag. 10

Ara

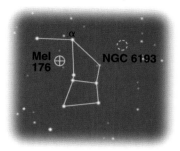

An ancient group showing the altar, named Thyterion by the ancient Greek astronomer Aratos of Soli, on which Centaurus the Centaur was to sacrifice Lupus the Wolf. Most of Ara sits within the Milky Way, which makes it jolly super to have a look at.

LATIN NAME
Ara
ENGLISH NAME
The Altar
ABBREVIATION
Ara
LATIN POSSESSIVE
Arae

α STAR
Ara
MAGNITUDE
3.0
STAR COLOUR
Bluish

The brightest star (actually two stars very close together) in the scene is the fifth-magnitude lead of the galactic cluster NGC 6193, with the wispy NGC 6188 seen flowing vertically towards the right.

EYE

DEEP SKY OBJECT
NGC 6193
TYPE
Galactic Cluster
MAGNITUDE
5.2
SIZE
15'
DISTANCE IN LIGHT-YEARS
3,750

There are around 30 stars here, a few of which can be resolved with binoculars. Larger telescopes will also show some faint nebulosity (NGC 6188) that fills the area on the cluster's western side. See the location chart in Norma, left, for this cluster.

BINOCULARS

DEEP SKY OBJECT
MEL 176
TYPE
Globular Cluster
MAGNITUDE
6.0
SIZE
26'
DISTANCE IN LIGHT-YEARS
7,200

A very fine globular – and one of the closest to us – which a small telescope will start resolving into stars. Some observers even claim to have seen this with the unaided eye on super-dark nights. Also catalogued as NGC 6397.

Lupus

LATIN NAME
Lupus
ENGLISH NAME
The Wolf
ABBREVIATION
Lup
LATIN POSSESSIVE
Lupi

α **STAR**
α Nor
MAGNITUDE
3.0
STAR COLOUR
Bluish

A king one minute, a wolf the next. And so the bloodthirsty monarch Lycaon of Arcadia became a howling constellation in the heavens. In some designs Lupus is held by Centaurus. Here is the Milky Way again – we just can't get away from it.

Mel 130

BINOCULARS

DEEP SKY OBJECT
MEL 130
TYPE
Galactic Cluster
MAGNITUDE
6.5
SIZE
40'
DISTANCE IN LIGHT-YEARS
3,000

A big loose collection of stars with no central clumpiness but some great star chains and gaps all over the shop, making it a great binocular object. Also designated as NGC 5822.

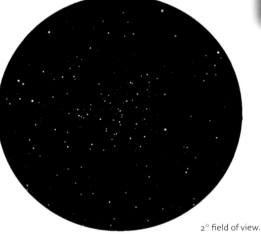

2° field of view.

Pavo

Designed by Frederick de Houtman and Pieter Dirkszoon Keyser, Pavo was included by Johann Bayer in his 1603 *Uranometria* star atlas. This is the bird that was sacred to Juno, goddess of the heavens – the one who was nasty to Callisto and changed her into a bear (see Ursa Major).

LATIN NAME
Pavo
ENGLISH NAME
The Peacock
ABBREVIATION
Pav
LATIN POSSESSIVE
Pavonis

α STAR
Peacock
MAGNITUDE
1.9
STAR COLOUR
Bluish

DEEP SKY OBJECT
MEL 218
TYPE
Globular Cluster
MAGNITUDE
5.4
SIZE
20.4'
DISTANCE IN LIGHT-YEARS
13,000

The best thing in Pavo – this fuzz is fairly easy to find because of its size, brightness and the fact that a seventh-magnitude star sits right in front of it. Also goes by the name of NGC 6752.

BINOCULARS

DOUBLE STAR
μ PAV
MAGNITUDE
5.3 & 5.7
SEPARATION
9'
COLOURS
Both orange

A wide pair that's easily seen if you have a dark sky.

Apus

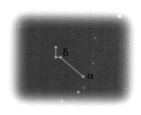

This is one of the designs that those Dutch mariners Frederick de Houtman and Pieter Dirkszoon Keyser made whilst sailing the southern oceans, but it has been known by many names through history. Eventually Apus found its way into Johann Bayer's *Uranometria* star atlas of 1603 and has been part of star lore ever since.

LATIN NAME
Apus
ENGLISH NAME
The Bird of Paradise
ABBREVIATION
Aps
LATIN POSSESSIVE
Apodis

α STAR
Aps
MAGNITUDE
3.8
STAR COLOUR
Orange

BINOCULARS

DOUBLE STAR
δ APS
MAGNITUDE
4.7 & 5.1
SEPARATION
1' 43"
COLOURS
Both red

October to December Skies

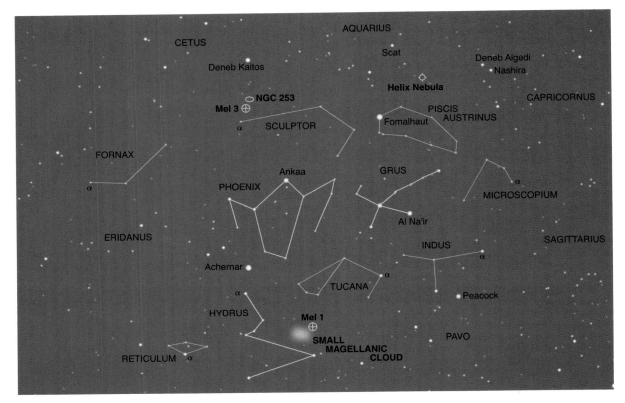

AQUARIUS

CETUS

Scat

Deneb Algedi

Deneb Kaitos

Nashira

Helix Nebula

CAPRICORNUS

NGC 253

PISCIS

Mel 3

AUSTRINUS

SCULPTOR

Fomalhaut

FORNAX

GRUS

Ankaa

PHOENIX

MICROSCOPIUM

ERIDANUS

Al Na'ir

INDUS

Achernar

SAGITTARIUS

Peacock

TUCANA

HYDRUS

Mel 1

SMALL

PAVO

MAGELLANIC

CLOUD

RETICULUM

- Galactic Cluster
- Globular Cluster
- Nebula
- Planetary Nebula
- Galaxy

The southern spring
skies looking south.

Eridanus

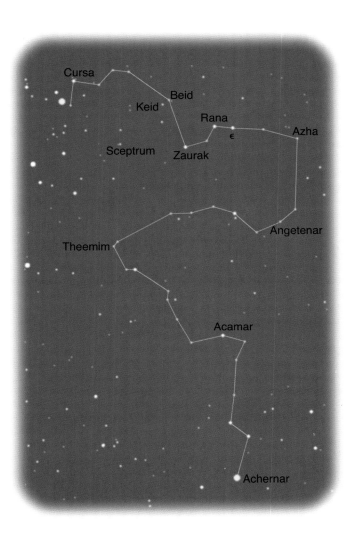

I'm not keen on chariot driving, especially at speed. This is what Phaeton found too, much to the disappointment of his father, Helios, the Sun god. Phaeton lost control and splash he went into the river – you should have seen the water damage, and it was only insured for third party, fire and theft.

Achernar must hold a record of superness in some space book somewhere for *'rotationallytastic squashiness whilst still keeping itself together'*, for it spins soooo fast that its equatorial diameter is twice that of the spin axis.

Epsilon (ε) is the third closest star to us after Rigel Kentaurus and Sirius. It sits just 10.7 light-years away and possibly has a system of planets.

LATIN NAME
Eridanus
ENGLISH NAME
The River
ABBREVIATION
Eri
LATIN POSSESSIVE
Eridani

α STAR
Achernar
MAGNITUDE
0.46
STAR COLOUR
Bluish

STAR
Achernar
DISTANCE
144 light-years
ABSOLUTE MAGNITUDE
-1.3
LUMINOSITY
2,900
SPECTRAL CLASS
B3

TELESCOPE

DOUBLE STAR
θ **ERI**
MAGNITUDE
3.4 & 4.5
SEPARATION
8.2″
COLOURS
Both white(ish)

Sculptor

LATIN NAME
Sculptor

ENGLISH NAME
The Sculptor

ABBREVIATION
Scl

LATIN POSSESSIVE
Sculptoris

α STAR
α Scl

MAGNITUDE
4.3

STAR COLOUR
Bluish

M11   Sobieski

Not one of Nicolas La Caille's best offerings. Originally he named it Apparatus Sculptoris, the Sculptor's Apparatus – an extravagant name to make up for the rather faint grouping of stars that actually look nothing like something a sculptor would use.

BINOCULARS

DEEP SKY OBJECT
NGC 253

TYPE
Spiral Galaxy

MAGNITUDE
7.2

SIZE
26'

DISTANCE IN LIGHT-YEARS
9 million

Notice that the size of this thing, sometimes called the **SILVER COIN GALAXY**, is nearly as much as the diameter of the Moon – zowie! Even though binoculars will only 'see' the central 20 arc minutes or so of this edge-on galaxy, it is still a great sight, appearing as a long fuzz. Telescopic people (that's not people who can extend themselves but those with a telescope) are pleasantly surprised by this bright galaxy's elongated shape – and why wouldn't they be? It was discovered by William Herschel's sister Caroline on 23 September 1783 and is probably the closest galaxy to us outside our Local Group.

The long NGC 253 galaxy is below with the roundy Mel 3 globular above. It's great when you have two deep-sky objects visually close together – it's like getting a two-for-one offer when you're observing.

DEEP SKY OBJECT
MEL 3

TYPE
Globular Cluster

MAGNITUDE
8.1

SIZE
14'

DISTANCE IN LIGHT-YEARS
28,700

This blob can be seen in the same binocular field as NGC 253 above, but requires more effort on your part due to its faintness. Others call this NGC 288.

NGC 253 up close. It's just like nearly sooo 'side-on' and all the splodgy gas and dust and stuff of the spiral arms are just like sooo 'there'. (Image courtesy Todd Boroson-NOAO-AURA-NSF)

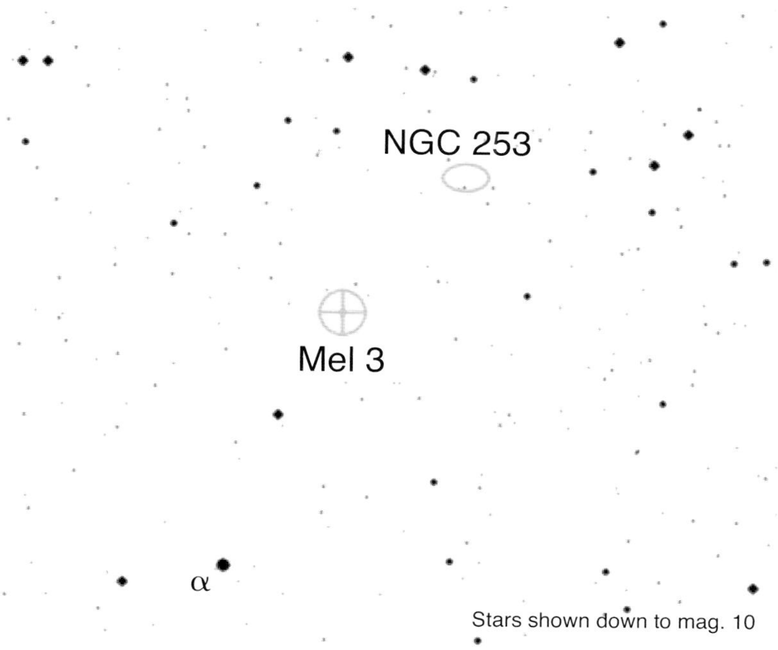
NGC 253
Mel 3
α
Stars shown down to mag. 10

Location chart for the things in Sculptor.

The star Fomalhaut leads the constellation Piscis Austrinus, the Southern Fish, that borders Sculptor. There is nothing of real note except this brightish star, and it had to go somewhere, so here it is. Hope that's acceptable to you all.

STAR
Fomalhaut (α PsA)

DISTANCE
22 light-years

APPARENT MAGNITUDE
1.16

ABSOLUTE MAGNITUDE
2.0

LUMINOSITY
13

SPECTRAL CLASS
A3

Tucana

LATIN NAME
Tucana
ENGLISH NAME
The Toucan
ABBREVIATION
Tuc
LATIN
POSSESSIVE
Tucanae

α Star
α Tuc
MAGNITUDE
2.9
STAR COLOUR
Orange

This is the South American bird with the long beak designed into the sky by Frederick de Houtman and Pieter Dirkszoon Keyser. That's about it really for the stars, while of course the main feature is the wonderful Small Magellanic Cloud (SMC).

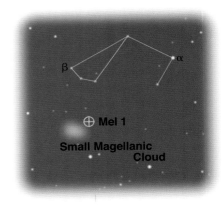

EYE

DEEP SKY OBJECT
SMALL MAGELLANIC CLOUD
TYPE
Irregular Galaxy
MAGNITUDE
2.3
SIZE
5°19' × 3°25'
DISTANCE IN LIGHT-YEARS
196,000

Here in splendid photographic detail, the Small Magellanic Cloud is the great big thing. To the right is Mel 1 with (and now I realize I'm in danger of sounding like a geek) another smaller globular, NGC 362, at the top.

Also catalogued as NGC 292, the SMC is one sixth the size of our Galaxy. This local little galaxy has been known through the ages, but particularly came to notice when Magellan sailed the world in 1519. To the eye it looks like a bit of the Milky Way that's broken off and is now wandering the starry skies. Actually it is wandering – very slowly: the SMC is orbiting our Galaxy as well as being gravitationally distorted by it. Sitting just next door is the smashing globular, Mel 1 and the double star β TUC, mentioned below.

DEEP SKY OBJECT
MEL 1
TYPE
Globular Galaxy
MAGNITUDE
4.0
SIZE
30'
DISTANCE IN LIGHT-YEARS
13,400

Old crazy astronomers thought this bright round blob to the right of the picture was a star – 47 Tucanae to be precise. Then, with modern-fangled telescopes, it became a big fuzzy blob the size of the full Moon! This is a brilliant binocular object, and it's a tough choice between this and Omega Centauri (see Centaurus) for the best globular in the sky. Also catalogued as NGC 104.

DOUBLE STAR
β Tuc
MAGNITUDE
4.5 & 4.8
SEPARATION
27"
COLOURS
Bluish & white

CHAPTER 8

The Moon and its Orbit

I'd like to make two points here, no, three (which are usually linked together, so it may actually only be two): the first one is that there's a lot of twaddle about the Earth and Moon being a double planet due in part to their small difference in size; secondly, some say that the Moon doesn't orbit the Earth, but actually orbits the Earth-Moon's centre of gravity; and thirdly, even more baloney that the Moon doesn't go round the Earth at all, but instead orbits the Sun. At this stage, don't get me started on the faked Moon landings! Some writers use all of the above points to verify the Earth-Moon double-planet status. So, one at a time now, and with a deep breath, I'll explain why this is

not true (remember, no planets or moons were harmed in the making of this chapter).

The Earth and Moon are *not* a double planet because the Moon orbits the Earth. The Moon is travelling quite happily in the gravity well of the Earth, completing each circuit in 27.3 days. The idea of a 'well' was used by Albert Einstein to explain how gravity works – and it's great as it isn't too difficult to visualize. At its simplest, imagine different weight people on a trampoline – the bigger the person, the larger the dip they make in the bouncy fabric. It works just like that in space, with small moons making smaller dips compared to larger planetary ones or the great dips made by stars. Of course we don't see these cosmic dips or wells, but we can imagine the idea that objects of different sizes curve the 'fabric of space'.

If you stand still on a trampoline and roll a marble around the outside of the dip you've

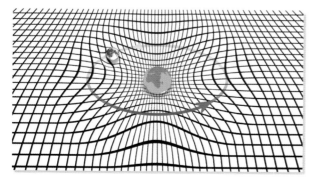

The gravitational dip of the Earth – it's caused by the mass of the Earth warping the fabric of space (and time too, if you want a deep discussion). The Moon rolls around the 'walls' of this well (notice that it makes its own dip – or warp if you like – too, but as the Moon is 81 times less massive than the Earth its dip is correspondingly smaller).

Here's my son, Morten, on his trampoline doing the Einstein thing. It's a simple demonstration of how setting a ball to roll within the dip made by Morten causes it to 'orbit' him – just as the Moon orbits the Earth.

if it orbited the Sun by itself then it would be a planet (it satisfies the size criterion which some people use) but by definition, as it orbits the Earth, it becomes a moon of the Earth.

'Ah,' they say, 'but because the Moon is large enough to be classed as a planet then it *should* be a planet.' Well then we go into the territory of some of the moons of Jupiter, Saturn and Neptune, which are also large enough to be classed as planets, and it all gets ridiculously muddled – why do some people try to complicate the whole solar system?

So let's just leave moons as they should be: moons.

'Ah,' they say, 'the Moon doesn't orbit the Earth – they both swing around their mutual centre of gravity.' This analogy is similar to sticking an object – one heavy, one light – on each end of a ruler and balancing the ruler on a finger: the balancing point (centre of gravity) is going to be closer to the heavier object. This is indeed what the Earth and Moon do, but their balancing point is within the Earth. If you imagine the Earth and the Moon as having exactly the same mass, they would orbit like two skaters holding hands as they spin – their centre of mass would be exactly in between. But as the Moon is so much smaller than the Earth, their mutual centre of spin is inside the Earth – i.e. the Moon 'goes around' the Earth.

'Ah,' they say, 'they have the lowest difference in size between any planet and its (largest) moon.' But again this is just some arbitrary choice of criteria. With any moon of

made, it'll do a sort of basic 'orbit' for a few seconds before the friction of the material causes it to slow down and fall headlong into the bottom of the dip. Thank goodness there is no friction in space – the Moon isn't actually on anything – or things would be crashing about like there was no tomorrow, and that would never do.

What we don't see from the Earth and Moon dip diagram is that they both sit within the great dip made by the Sun – as does everything in the solar system – so the Earth is just doing what the Moon does but on a much larger scale.

Imagine now removing the Earth from the scene. What would the Moon do? (ponder, ponder…) Free from the Earth's gravitational attraction, it would follow whatever gravitational well there happened to be curving around, and hence it would begin orbiting the Sun. Now, the 3,476km diameter of the Moon does mean that any planet you could argue it should not be a moon for some reason: because of orbit, shape, size, origin, etc. What do you fancy choosing today? So, the claim that, because the Earth is 'only' 50 times bigger than the Moon, this makes them a double planet doesn't hold with me.

Finally there's the 'change in viewpoint' challenge. As seen from the Sun, both the Earth and Moon seem to move together in the same direction in the sky. You might think that because the Moon orbits the Earth, the Moon would be seen to move ahead and then swing back over the month, but no. The Earth is seen to travel at a more constant rate, while the Moon is seen to speed up and slow down, but never to change direction. Imagine being in a dinghy on a safe but fast-flowing river. The dinghy represents the moving Earth (it's just being taken along with the current), you are the Moon, and a friend with a towel represents the Sun (and its viewpoint), sitting on the riverbank, waiting excitedly for your demonstration.

'I'm ready,' you cry to your friend, and jump out of your dry craft into the fast-flowing alpine glacier melt-water. Catching your breath with a silent 'It's f-f-freezing!', you swim downstream, easily moving ahead of the dinghy – this is equivalent to the Moon moving in the same direction as the Earth through space. After 7.375 days you turn around and swim upstream against the current (okay, you don't need to be so precise unless you want to do it properly). You eventually catch up and move past the dinghy, but the current is so strong that, as far as your friend, the Sun, is concerned, you are still moving downstream, just not as fast. Say the river is flowing at 4kph, but you can only swim at 3kph – if you swim with the current you're doing 7kph downstream; swim against the flow and you're still going downstream, but only at 1kph relative to the bank. As seen from the Sun, therefore, this is what the Moon does as it orbits the Earth: half a month with the flow and half a month against, but always seen moving in the same direction, albeit at differing speeds.

In space this comes about because the orbital speed of the Earth is 30 times faster

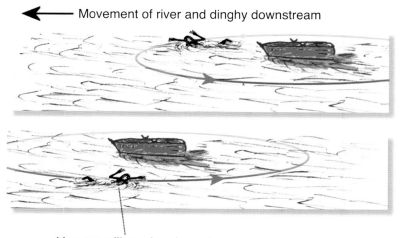

Movement of river and dinghy downstream

I know I'm not the world's greatest artist, but here is a watery artistic demonstration of the Earth (the dinghy) and Moon (you) in orbit (red circle) as seen from the Sun (the riverbank). Top: You are swimming with the current and appear to race ahead downstream. Bottom: Because of the flow of the river you are at no time ever seen to move to the right – even when you are swimming against the current your speed is too slow to make any progress back up the river.

You are still moving downstream as your speed is not fast enough to beat the current than the Moon's. Our planet is racing around the Sun at 28.7km/s (mean), while the sedately travelling 1km/s Moon has little chance of making an impact. But to suggest that this movement as observed from the Sun (often described as *co-orbital*) is a further example of double-planetary status is simply convenient for the argument. If we use another viewpoint, say the centre of the Galaxy, then relative to the distant galaxies the Earth and Sun are seen to move co-orbitally as well. Therefore I could say the Earth doesn't go around the Sun at all. So if you ever meet one of those double-planet people, tell them to go and get a dinghy.

All done. Quite happy? Now we can get down to the nitty-gritty stuff: how does the movement of the Moon affect how we see it in the sky? The diagram here gives a closer-to-reality view than in many books where both objects are shown large and too close together. Yes, that little black dot is the size of the Moon compared to the Earth (although I have a sneaking suspicion I've made the Moon a little too big!). It's quite surprising to see how far away the Moon really is from the Earth: the mean distance is 384,000km – that's 30 times the Earth's diameter.

Our diagram viewpoint is from far above the North Pole. If we watched over time from there, it would become apparent that the Moon orbits the Earth anticlockwise. Taking the dot as its starting position and rolling time forwards, we would see the Moon travel around us, returning to the same place after 27.32 days – this length of time is known as the **sidereal month**.

This is also the time it takes for the Moon to spin exactly once on its axis. However it is not the time from one Full Moon to the next. If you

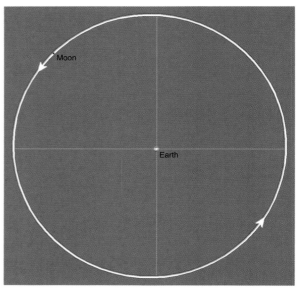

The size and spacing of the Earth and Moon. It's all as perfect as I can make it without wearing an anorak. (Orbit plotted by Scientific Astronomer software, Wolfram Research, Inc., Mathematica, Version 5.2, Champaign, IL [2005])

are out one clear night viewing a marvellous fully illuminated lunar globe, you have to wait 29.53 days until the next one – a length of time known as the **synodic month**. (You may remember we talked about **sidereal** and **synodic days** in Chapter 1, with reference to the spinning Earth.)

The time difference comes from the fact that the Earth is orbiting the Sun and because of that,

The far side of the Moon as seen by Apollo 16 in April 1972. We can never see this view from Earth as the Moon spins on its axis in the same time as it orbits the Earth, thereby always keeping this side facing away from us.

over the month, the Sun appears to have moved across the sky, also anticlockwise, just a little bit. If you're going to move the Moon's main source of illumination, then you cannot expect much else, to be honest. The Moon must 'catch up' and, just over two days later, it's back to its facing-the-Sun position again: Full Moon.

The cross hairs on the diagram show how the Moon's orbit is not centred on the Earth. The orbit is not a circle either, but has an oval or **elliptical** nature. The more elliptical an orbit, the higher the *eccentricity* it is said to have. Whereas an object with a circular orbit will keep the same distance from the body it is travelling around, anything with an elliptical orbit will have a point when it is at its closest and another when it is furthest away.

The instant the Moon is closest to the Earth is called *perigee* (the blue arrowed line in the diagram above right), when it is at present 363,104km away. *Apogee* represents the furthest point in the Moon's orbit, currently standing at 405,696km (the red arrowed line). So, one take away the other leaves a difference of 42,592km.

Not surprisingly, this leads to the Moon appearing larger or smaller in the sky over the course of the month. The picture of the two Full Moons next to each other reveals the observed difference in apparent size between perigee and apogee Moon. The difference is easiest to notice with Full Moons but, remember, as the Earth travels along its orbit, these points move in relation to the Sun, so apogee and perigee can and will happen at any phase, be it waxing or waning crescent, half, gibbous or full.

As for the Full Moons, it is true you can notice the difference in size when they fictitiously sit

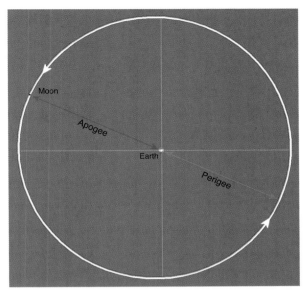

The ellipticity (how elliptical or oval something is) of the Moon's orbit means that during the month there is an occasion when the Moon is closest to the Earth – this is perigee, while the point when it's furthest from the Earth is apogee. (Orbit plotted by Scientific Astronomer software, Wolfram Research, Inc., Mathematica, Version 5.2, Champaign, IL [2005])

next to each other, but in the real sky this is much more problematic for a variety of reasons. Firstly, it is quite difficult separately to gauge such small differences, especially given that a Full apogee Moon happens six months after a Full perigee Moon. It's another one of those astronomical challenges to try. And don't forget, it *will* be cloudy on that important day in question when you have geared yourself up to prove that this is possible.

The difference in Full Moon sizes at perigee (left) and apogee.

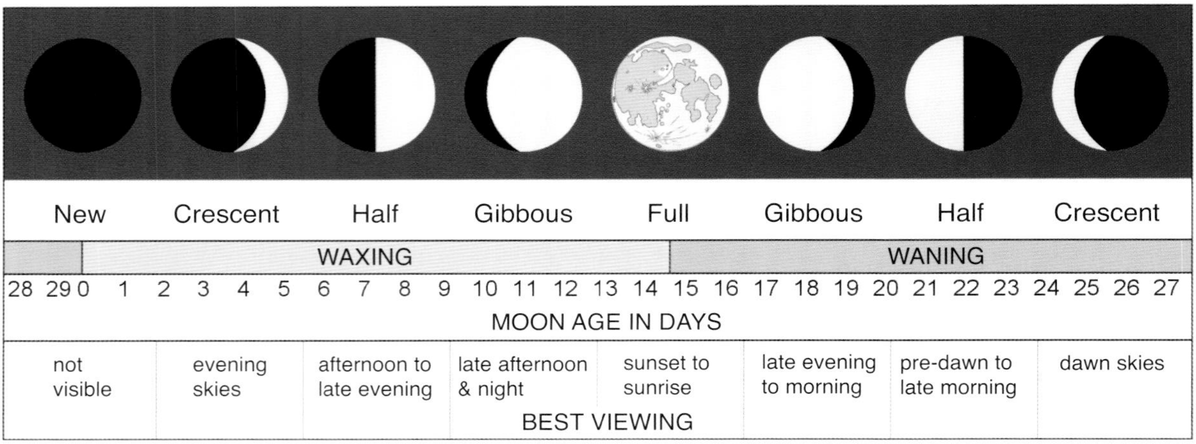

	New	Crescent	Half	Gibbous	Full	Gibbous	Half	Crescent
		WAXING				WANING		
Moon age in days	28 29 0	1 2 3 4 5	6 7 8 9	10 11 12 13 14	15 16	17 18 19 20	21 22 23	24 25 26 27
Best viewing	not visible	evening skies	afternoon to late evening	late afternoon & night	sunset to sunrise	late evening to morning	pre-dawn to late morning	dawn skies

Chart of the Moon's phases over the 29.53-day synodic month.

Observing the Moon

The telescopic view of sunrise or sunset on parts of the Moon, when features are illuminated by low sunlight casting dramatic shadows across the lunar surface, is incredible. Think of the way the light from the setting Sun on Earth creates long shadows that reveal the relief of the landscape, and you'll see what I mean. This boundary of light and dark is called the **terminator** and it's around at any time except New and Full Moon. Some of the Moon's features are large enough (and rough enough) that we can see them with the unaided eye if they are caught on or near the terminator.

Look a little longer than a glance at the Moon and other details become apparent on its rocky surface. The most prominent of these are *craters*, such as Copernicus, that are noticeable from the 'splat' of brightness around them, which formed as the crater-creator object (be that a piece of asteroid, of a comet, a small fridge, etc.) smashed into the Moon, sending up a mass of stuff that was thrown out and deposited across the lunar surface.

There are many other things to take a look at as well, some of the most popular of which are described here. So, grab a cup of tea and a sandwich. Find those dusty binoculars (not that you'll need them to see everything) and prepare for more adventure as we count down our Moony Top Eight highlights, or, as I like to call it, Now that's what I call eight Moony features:

Lunar Features

Seas (Latin singular *mare*; plural *maria*)

These are those large dark patches that are the most notable eye features, occurring mainly on the Earth-facing side of the Moon. They were formed when big things hit the younger Moon, causing cracks in the surface from which molten rock oozed out to fill in any large basins that happened to be nearby. The result was also that

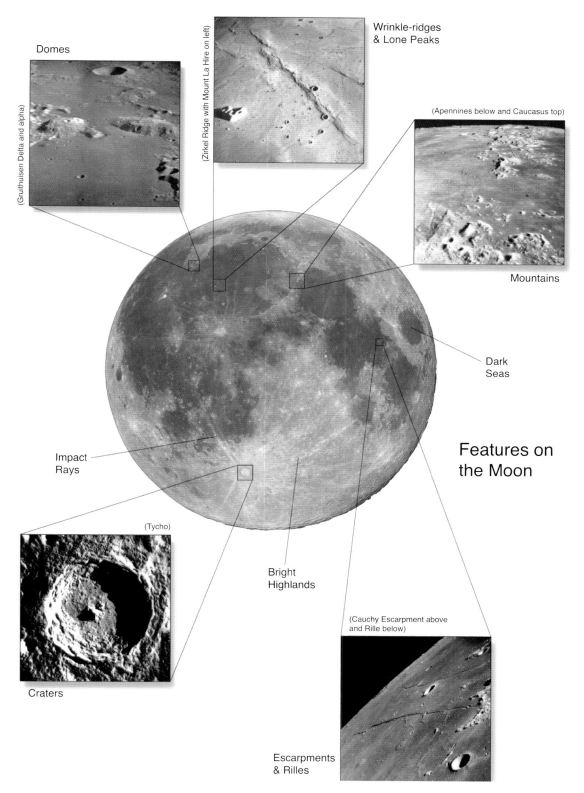

Domes

(Gruithuisen Delta and alpha)

(Zirkel Ridge with Mount La Hire on left)

Wrinkle-ridges
& Lone Peaks

(Apennines below and Caucasus top)

Mountains

Dark
Seas

Features on
the Moon

Impact
Rays

(Tycho)

Craters

Bright
Highlands

(Cauchy Escarpment above
and Rille below)

Escarpments
& Rilles

the older cratered surface was 'iced over', as it were, leaving us what we see today: a relatively flat and fairly clean-looking surface. Ancient astronomers without telescopes took these darkened areas to be water, hence the misnomer 'seas'. Depending on their size, there are also bays (Latin singular *sinus*), marshes (Latin singular *palus*), lakes (Latin singular *lacus*) and one ocean (Latin singular *oceanus*).

Biggest watery feature: Oceanus Procellarum
(Ocean of Storms)
at 2,578km in diameter
Best watery name: Lacus Perseverantiae
(Lake of Perseverance)
Worst watery name: Palus Putredinis (Marsh of Rot)
Best general seas viewing: around Full Moon
Minimum apparatus needed: eye

Mountains (Latin singular *mons*; plural *montes*)

These come singly or in chains, many with Earthly names – for example, there's Mont Blanc in the Alps (the only mountain on the Moon not spelt 'mons'). Everywhere on the Moon, away from the dark seas, are the *highlands,* with mountains of some sort or other. Some notable ranges occur around the edges of the seas, while the odd peak, too high to be covered by the lava when the seas formed, stands solitary in an otherwise desolate flat landscape. Unlike the Earth, there is no rain, wind, frost or anything to wear away the lunar mountains, so they stay rough and science-fictiony jaggedy, with heights nearing 5km.

Highest mountain: Mons Huygens, found in
Montes Apenninus, at 4.7km
Longest mountain range: Montes Rook at
791km
Best viewing: close to the terminator
Minimum apparatus needed: binoculars

Wrinkle ridges (Latin singular *dorsa*; plural *dorsum*)

If it's a gently winding low ridge (like a long, snaky hill) found on the seas, sometimes stretching for hundreds of kilometres, then it's a wrinkle ridge. These were formed as the lava that made the seas cooled, contracted, moved along faults and/or sagged under its own weight, sometimes giving a hint of now hidden features that lie underneath. As they are no more than a few hundred metres in height, the lower the sunlight that catches them, the bigger the shadow cast and hence the easier it is to see them well.

The raised feature snaking up the sea from the lower left is the famous Smirnov Ridge (Dorsa Smirnov), ending up close to the full-of-structures-and-things crater Posidonuis. (Image courtesy NASA)

Rilles (Latin singular *rima*; plural *rimae*)

As opposed to ridges, which rise up from the surface, rilles are those things that cut into it like a groove or trench. They are quite wide, generally stretching over a few kilometres, with lengths up to several hundred kilometres. They were made where the land had sunk for some reason or another – maybe the roof of a tube that used to carry lava, or the land between two faults, collapsed.

Domes

These are, just as their name suggests, dome-like features, believed to have been made by volcanic activity and found dotted around the lunar seas. Sticky lava erupting slowly from something like the shield volcanoes we have on Earth took its time to cool into structures just a few hundred metres high by up to 20km across and sometimes with a small crater on top. Other domes may have been a consequence of rock being pushed up by a growing magma chamber below the surface.

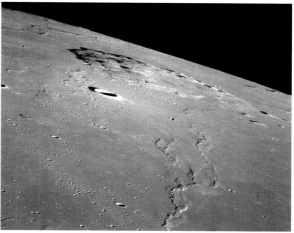

LEFT: Sunrise over any landscape defines its features marvellously with highlights and shadows. Here is the area around the twin craters Feuillée and Beer (top), with wrinkle ridges, rilles, lone mountains and other craters easily picked out by the low sunlight. (Image courtesy NASA)

ABOVE: A collection of volcanic domes has produced the mounds known as the Rumker Hills (Mons Rumker). (Image courtesy NASA)

Escarpments (Latin singular *rupes*)

These are straight features caused by a long vertical shift of the crust along a fault. Compression forces the land on one side of the fault up (or the other side down) and a cliff face, at a variety of angles, results.

> *Longest escarpment:* Rupes Altai at 427km
> *Best viewing:* close to the terminator
> *Minimum apparatus needed:* binoculars

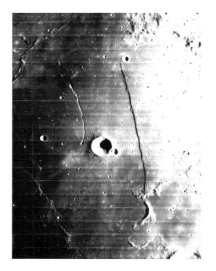

The famous Straight Wall (Rupes Recta) escarpment in Mare Nubium, sitting to the right of the happy-go-lucky crater Birt. There is also a fine rille running north just off to the left of the crater. This is a waxing Moon view, when the escarpment face is in shadow. During the waning phase the wall shows up bright, as it is illuminated by the Sun. (Image courtesy NASA)

The Schroter Valley (Vallis Schröteri) cuts its way down from the middle of the picture, then curves off and back up to the right, in the great landscape dominated by the big bright crater in the upper left, Aristarchus. This view is from the Apollo 15 lunar module, hence not as you would expect to see it unless you have a lunar module yourself and plan to take this exact trip. (Image courtesy NASA)

Rays

Smash something (like some size of rock or meteoroid) into something else (like the Moon) at incredible speeds and you get an explosion. Due to the lower gravitational pull of the Moon, the resulting debris (or *ejecta*) thrown up and away from the explosion spot (the crater in this case) takes longer to come down and so travels further than it would on Earth. Various objects at various speeds have hit the Moon over the eons and left some fine bright trails of debris across the surface – these are the rays, all emanating like a paint splat from the impact crater. You can create the same effect by throwing eggs really hard at the floor – watch and enjoy the resulting rays (please wear protective clothing and goggles, and don't forget to remove small pets and children from the impact and ejecta zone).

> *Longest rays:* 1,500km long from crater Tycho
> *Best viewing:* around Full Moon
> *Minimum apparatus needed:* eye

And the winner is...

Impact Craters

What else could win other than the features that epitomize the Moon? Craters, hundreds of thousands of them, pockmark the lunar surface, with many more buried beneath various lava flows and the Moon's powdery-rocky surface (or *regolith*). As I mentioned under Rays, above, if something smashes into the Moon, there is an explosion. This melts some or all of the meteoroid and the surface of the Moon. The molten material travels outwards in a circle(ish), which – depending on the force and loads of other space whatever – reaches a point of resistance that forms the outer rim of the crater. If there's enough energy and the rock is still molten enough it can bounce back to meet in the middle where the energy collides to form a peak of lava – that must be an amazing thing to see being made. With any luck it solidifies here and we get a crater peak.

After you've cleaned up the eggs from the 'ray' nonsense above, move over to the sink and fill it with water, then let single drops fall into it. The drop is akin to the meteoroid, while the ripples it causes are like the molten rock forming the crater. I like to think of a crater as a frozen ripple.

Not all craters have peaks and, with some, lava has filled and flattened the floor of a crater, in which case they are known as a *walled plain*. Other effects due to gravity have caused the outer walls of some craters to slump into a terraced appearance. And one other thing: you only get nice circular craters if the object hits the Moon from directly above and there is nothing else around to affect the molten flow. Therefore we see many oval and occasional odd-shaped craters.

There are also craters, as we have seen with domes, that are of volcanic origin, and there are hints that there is still some volcanic activity on the Moon today – not in the form of big events, but in a sporadic, low-key, quiet, whispery and mysterious kind of way. Occasionally observers have noted mists, colourization or glows in some of the craters, and all these types of things have become known as *Transient Lunar*

The crater Plato on the left, with the Alpine Valley scarring the Alps on the right. The wonderful mountain on its own in the sea to the lower right is the 2.3km-high Mons Piton. The other lone peak directly below Plato is 2km high and called Mount Pico.

Phenomena, or *TLP*. They are probably due to the release of gases, but they are so few and far between that a full explanation of all the effects could be a long way off. Needless to say, if you see some ghostly, wispy, glowy thing hanging around one of the craters indicated in the table below, then you too could be one of the very few who have seen a TLP. That, or you really need to clean your eyepieces.

Biggest crater: Bailly at 295km in diameter
Deepest crater: Newton at 8.8km from rim to base
Best viewing: it depends on the crater, but close to the terminator works for them all
Minimum apparatus needed: binoculars

Lunar Craters

The table opposite is a run-down of the craters featured in the picture. The column on the right gives you a rough indication of the time when the light from the rising Sun catches the crater at a good low angle for you to see it at its best – shadows and relief and all that. Not all craters are at their best at the Full Moon, due to the lack of shadows, but for some it is an ideal time to look, hence the word 'Full' accompanying some of the days.

I have only given the days for a waxing Moon – that is from New to Full Moon when the sunrise (and therefore the terminator) sweeps leftward across the lunar surface. Of course after Full Moon the sunset does exactly the same and catches all the light, which now hits the craters from the opposite angle – it's incredible how this change of light direction can cause most craters to look so different. To get the time when the waning Moon sunlight catches the crater, add 14 onto the day number given.

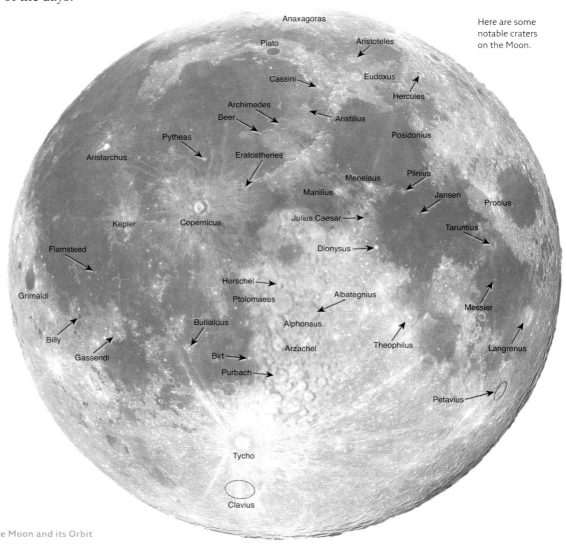

Here are some notable craters on the Moon.

Crater	Diameter (km)	Description	Days after New Moon to get a good view
Albategnius	129	A large, squarishly shaped old crater sitting in the lunar highlands with terracing and an elongated central peak. The crater Klein overlays the lower-left rim.	7
Alphonsus	119	This large crater just cuts into Ptolomaeus above it. The floor contains rilles and a central low peak.	7
		Alphonsus is one of the places where TLP have been seen.	8 & Full
Anaxagoras	51	Right at the top of the Moon, this young crater stands out due to its bright rays, some of which stretch 600km.	
Archimedes	83	A great, smooth, walled-plain, dark-floored crater. Off to the left are the small twin craters Beer and Feuillée, while the Marsh of Decay with its system of rilles sits to the lower right.	7
Aristarchus	40	This is considered the brightest natural feature on the Moon. The steep central peak is bright, so are the outer terraced walls and so is the ejecta ray system.	10 & Full
Aristillus	55	A cluster of peaks lies in the centre, while material thrown out during the formation of the crater can be seen on the Mare around the outside.	7
Aristoteles	87	Here's a crater with high walls that rise over 3km from the floor. The outer walls themselves are terraced, and the small crater on the lower right rim is called Mitchell.	5–6
Arzachel	96	High crater walls, a jaggedy central peak and a few rilles make this youngish crater well worth a telescopic viewing.	7
Beer	10	One of two small but very fine identical twin craters – its brother, Feuillée, is the top left one.	9
Billy	45	This crater has been deeply flooded by lava, leaving a featureless dark floor.	11
Birt	17	Bowl-shaped Birt sits just to the left of the famous 'Straight Wall'. The smaller crater that sits on the lower-right rim is the inspirationally named Birt-A.	8
Bullialdus	60	A great crater with lots going on. It firstly appears notable because it and its 'splat' sit in the dark Mare Nubium. Bullialdus has several high central peaks, with internal terraced walls and a bumpy undulating floor.	9
Cassini	57	A somewhat irregular walled plain with low walls and several smaller craters on the floor.	
Clavius	225	This is a biggy – a great irregular ancient depression with 3.5km-high walls and smaller craters through the area. Have a look with the eye at the Moon when it's eight or nine days old and the feature you notice near the bottom of the terminator is Clavius.	8–9

Crater	Diameter (km)	Description	Days after New Moon to get a good view
Copernicus	95	Great crater and splat on the dark Oceanus Procellarum; multiple central peaks make it good for viewing at Full Moon too.	9 & Full
Dionysus	18	A very bright little crater with a small ray system. There is a rille system just off to the top left.	6 & Full
Eratosthenes	58	A great 3.6km-deep impact crater with high central peaks and internal terracing. It sits at the lower end of the Apennine Mountains and is partially covered by the rays from Copernicus.	8
Eudoxus	70	A great step-sided crater sitting close to Aristoteles.	6
Flamsteed	20	Small with quite steep and high walls and sitting in the lower part of a large 'ghost' crater (known as Flamsteed P) that was flooded during the creation of Oceanus Procellarum.	10
Gassendi	110	A fine crater with a lean – its southerly walls are only a few hundred metres high, while to the north they are approaching 2.5km. There are multiple peaks poking out of the flooded floor which contains rilles galore, and it's a place of TLP.	10
Grimaldi	220	This is a large, old and so heavily 'attacked' basin that its walls have been pounded down into hills and mountains. It holds the title as the darkest feature on the Moon due to the flat flooded floor. TLP have been seen here.	13 & Full
Hercules	69	A 3.3km-deep crater which was lava-flooded, giving it a darkened appearance. The central peak was also covered during this time, leaving only a low central hill.	3-4
Herschel	42	This is the youngish circular crater with terraced inner walls and a low central peak that sits directly above Ptolomaeus.	7
Jansen	23	Small, but prominent due to its position close to the middle of Mare Tranquillitatis. Its floor is lava-flooded and there are rilles and wrinkle ridges nearby.	6
Julius Caesar	90	A lowish, irregularly shaped, partially submerged crater with darkening on its floor that flows out into the nearby Mare Tranqullitatis.	6
Kepler	32	Found in the centre of the bright smaller splat in Oceanus Procellarum to the left of Copernicus, this young crater has a ray system stretching over 300km.	10 & Full
Langrenus	132	A couple of central peaks, massive terraced walls rising 2.7km and some fine rays make this a substantial crater.	2-3
Manilius	38	The bright walls mean this crater stands out very nicely around Full Moon. It also has a central peak and a system of rays covering 300km.	6 & Full
Menelaus	27	A bright young crater to be found near the end of the Haemus Mountains at the bottom of Mare Serenitatis.	7 & Full

Crater	Diameter (km)	Description	Days after New Moon to get a good view
Messier	11	A tiny thing – great, though, as it's one of a twin, plus there is an unusual spray of ray-ish material that looks a bit like a comet whooshing off to the left.	4
Petavius	178	Catch it when the Moon's young and this superb crater reveals its central peaks and a straight rille that points like a clock-hand to 8pm.	2–3
Plato	101	A great large impact basin with a flat, dark grey floor. It is perfectly circular but appears oval due to the curving surface of the Moon away from us. There are also some tiny craters on Plato's floor and this has been a site of TLP.	8
Plinius	43	To be found in the small white splat between Mare Tranquillitatis and Mare Serenitatis, Plinius has an oval appearance with an undulating floor and terraced inner walls.	6 & Full
Posidonius	96	This fairly large structure, with a rough and bumpy lava-flow interior, sits on the coast of Mare Serenitatis, with some fine wrinkle ridges just offshore. The crater floor has many features, including rilles.	5
Proclus	29	It's bright all right, in fact the second brightest feature after Aristarchus, and it made a pretty good splat of rays to the left of Mare Crisium.	4
Ptolomaeus	153	A large, ancient walled plain that sits just below the visible centre of the Moon. As well as a few small craters on the lava-covered surface, the most notable of which is Ammonius to the top right, there are hints of craters that now lie buried underneath.	7
Purbach	118	A largish old irregular crater with walls that not surprisingly show signs of age.	7
Pytheas	20	This is the bright dot in Oceanus Procellarum just above Copernicus. A wrinkle ridge runs south from here to two small twin-like craters called Draper and Draper C.	9 & Full
Taruntius	56	This nice crater, outlined by its walls and darkish floor, sits between Mare Tranquillitatis and Mare Foecunditatis. The crater Cameron cuts across the top-left part of the crater rim.	4 & Full
Theophilus	101	One of the finest lunar craters, with high bright central peaks and collapsed terraced walls – great with the Sun at a low angle.	5
Tycho	84	Tycho has inner terraced walls, a main central peak with a smaller one to the north and a rough floor. However, what stands out with the eye is the greatest set of rays on the Moon. Bright 'spokes' of ejecta emanate from this crater and stretch up across the Moon for over 1,500km – that must have been one doozy of a collision to witness 108 million years ago!	9

Other Lunar Features

Sinus Roris
Jura Mountains
Sinus Iridum
Rumker Hills
Tenerife Mountains
Alpine Valley
Mare Frigoris
Alps
Caucasus Mts
Lacus Somniorum
Mare Imbrium
Mount Piton
Smirnov Ridge
Schroter's Valley
Mare Serenitatis
15
17
Oceanus Procellarum
Carpathian Mountains
Apennine Mountains
Haemus Mts
Mare Crisium
Sinus Aestuum
Mare Vaporum
Mare Tranqullitatis
Sinus Medii
Arago Alpha & Beta
12
14
Ariadaeus and Hyginus Rilles
11
Mare Foecunditatis
Pyrenees Mountains
16
Sirsalis Rille
Mare Humorum
Mare Numbium
Mare Nectaris
Hippalus Rilles
Straight Wall
Rheita Valley

● Apollo landing sites

Apart from the Apollo lunar modules, which are far too small to be seen from the Earth, have a go at some of these other features with a telescope.

Feature	Latin Name	Some value of width or length etc. (kms)	Description	Days after New Moon to get a good view
Alpine Valley	Vallis Alpes	180km long	Great rift fault across the Alps with a width varying between a few and nearly 20km.	7
Alps	Montes Alpes	560 x 240km	2.4km high, with many peaks along the Imbrium coast, including Mont Blanc at 3.6km high.	7
Apennine Mountains	Montes Apenninus	600km range	5.4km at its highest, with steep, 30-degree angle slopes, the range forms the lower-right wall of Mare Imbrium.	7
Arago Alpha & Beta	Arago Aplha & Beta	each 20km in diameter	Two volcanic domes, one above and one to the left of the crater Arago indicated.	5
Ariadaeus Rille	Rima Ariadaeus	225km long	Long trench, partly flooded in places, stretching from Mare Tranquillitatis to Vaporum with a branch down to join Hyginus Rille.	6
Carpathian Mountains	Montes Carpatus	340 x 100km	Some great rugged mountains separated by large valleys.	9
Caucasus Mountains	Montes Caucasus	500 x 120km	3.6km high. A varied mountain range that sinks in the south into Maria Serenitatis and Imbrium.	6
Haemus Mountains	Montes Haemus	411km range	Southern-left border of Mare Serenitatis.	6
Hippalus Rilles	Rimae Hippalus	240km long	Large parallel rilles running vertically on the right side of Mare Humorum.	9
Hyginus Rille	Rima Hyginus	220km long	Long trench that connects to the Ariadaeus Rille.	6
Jura Mountains	Montes Jura	480km range	6km-high, north-curved wall of Sinus Iridum basin; several concentric sets of peaks.	10
Mount Piton	Mons Piton	25km diameter	Height of 2.3km, a lone peak on a sea – great shadow.	7
Pyrenees Mountains	Montes Pyrenaeus	260 x 70km	Forms the right-hand border of Mare Nectaris.	3
Rumker Hills	Mons Rumker	70km diameter	A build-up of lunar domes forming a circular bumpy structure up to 500m high.	12

Feature	Latin Name	Some value of width or length etc. (kms)	Description	Days after New Moon to get a good view
Rheita Valley	Vallis Rheita	370km long	A great valley actually formed by several overlapping craters that have eroded over time.	4
Sirsalis Rille	Rima Sirsalis	425km long	This holds the record as the longest rille on the Moon and is unusual in that it only runs across the highlands.	12
Schroter's Valley	Vallis Schröteri	165km long	This begins as the widest rille on the Moon at 10km. It flows north, then turns to the left, getting thinner as it goes, so that by the end it is only 500m wide.	11
Smirnov Ridge	Dorsa Smirnov	156km long	A great system of wide wrinkle ridges running north–south on Mare Serenitatis.	5
Straight Wall	Rupes Recta	115km long	This fault has a typical width of about 2.5km and a height of around 250m. So, even though it looks like a vertical cliff in the lunar surface, the slope is not that steep.	8
Tenerife Mountains	Montes Teneriffe	105km range	A great submerged-looking range, with six main peaks rising 2.4km above Mare Imbrium.	8

CHAPTER 9

Moon and Sun Illusions

Imagine one of those glorious evenings, the sort that impresses us no end. Over there in the west the Sun has just set, leaving the horizon awash with hues of red, orange, pink, purple, smouldering violet, etc. As you turn around to pick up your glass of orange juice your gaze falls upon the other side of the sky, where the Moon is rising majestically from behind the tall pines. With the warm, late-summer breeze the scent of the trees fills the air. What an incredible Moon, you think: it's tinged with red and looks so large.

You take a seat to watch as the scene unfolds over the next few hours. Oh no, you drifted off. When you find the Moon again it has moved high into the sky, but it looks different – somewhat smaller. What has happened? Can the Moon have moved that much further away during such a short time? Maybe there is a more sinister reason, possibly a government cover-up.

The truth lies in a phenomenon called the *moon illusion.* Just as with any other optical illusion, our eyes and brain are playing tricks with what we think we see. There are many different discussions on the exact cause, indeed several bits of mind magic maybe joining forces to make sure we are well and truly fooled, but, for me, there is one that sticks out a mile – enter the Ebbinghaus illusion.

The Ebbinghaus illusion. No, surely it can't be true!

The trickery is shown here with the diagram of red circles surrounded by blue circles. Here's the thing: what do you notice about the two red circles? To me (and to all those poor souls I inflicted my artwork on before this went to print) the lower-left red circle (which represents the low Moon) looks larger than the upper-right one (the high Moon). In fact both the red circles are exactly the same size. I know that some of you are going to check, but for those of you without a ruler, I can assure you it is true. Honest.

What's happening is that the blue circles are affecting how we perceive the reds. Even when you know what's going on the brain-bending thing still works. To quote Monty Python, 'The illusion is complete.'

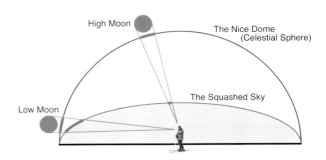

High Moon

The Nice Dome
(Celestial Sphere)

Low Moon

The Squashed Sky

I'm standing at the top of Run 1 in the Dachstein-West ski area of Austria, but at −18°C I shall not be hanging around for long. The basic message is that the sky above me (and you) looks squashed and not a Nice Dome shape. The green lines represent the size of the Moon on the Nice Dome – notice that they are the same length whether the Moon is high or low. However, when the Moon is projected down to our perceived Squashed Sky, the purple lines tell a different story – clearly smaller when high up. The reason for this is that my brain is having fun with an effect called size constancy and that Ebbinghaus illusion.

So, exactly how does this work with the sky? Let's consider me skiing for a moment. You need at least some of your horizon to be clear to really see this effect (and on not too cloudy a day either). Here on the Hornspitz mountain my entire horizon is filled with distant snow-capped peaks and in some directions I'm sure I can see 80-odd kilometres, if not more.

What I notice if I look all around and up and down is that the sky does not appear everywhere as equally distant: the sky above me actually seems closer than the sky near the horizon.

The effect is particularly easy to see when there are some friendly clouds hanging about the place, as it looks as if these puffs of moisture get further away as I gaze from those directly above my mountain to those in the distance. 'Well of course they do!' I hear you cry. Indeed they do, and that is the crux of the illusion. Even without the clouds it still appears that the sky directly overhead is close and as we look towards the horizon the sky gets further away.

This means that instead of every part of the sky we look at appearing to be at the same distance away from us, the higher up we look the closer the sky seems to be – almost as if it is squashed down on us.

The changing distance of the squashed sky brings into play an effect called *size constancy*. We see this all the time, as it's our way of judging size against distance. Take Kevin and

Clouds over Salt Lake City travel a long way out of town as they head towards my horizon. This gives the sky an appearance of being closer above my head than near the horizon. And guess what? It works without the clouds too – giving the sky a somewhat squashed-down-above-me look.

me skiing down the mountain. I am clearly much the better skier, so with you waiting at the bottom with some nice warm drinks for us, you see me arriving first and Kevin struggling far behind. Even though we are in reality about the same physical size, he appears smaller, of course, because he is further away. Nothing strange there.

So just trust me when I say that the Moon is the same size wherever it is in the sky – high or low. The problem with the Moon is that it is big and a long way away. As I explained in my last book, *Simple Stargazing*, trying to imagine sizes and distances in space is nigh on impossible, because they are so mind-bogglingly huge, so unlike the distance and size between Kevin and me on the mountain, which is a breeze for us to work out, the Moon gives us a challenge as there is nothing around it apart from the sky. Therefore the sky itself forms our only visual reference, and due to our perception of the far horizon sky and the close overhead sky, this is where the illusion starts.

Now look back at Ebbinghaus. Notice those blue circles. Size matters. If the sky was the Nice Dome, then you can imagine that those blue circles would look the same size everywhere. However, with the Squashed Sky, the circles above your head look bigger – because they seem closer – than the ones near the horizon. Therefore, and with a big fanfare, when the Moon is low, it is seen against a more distant, 'smaller' sky (small blue circles) and looks big. Conversely when the Moon is higher it sits against a closer 'bigger' sky and looks smaller.

Thank you and goodnight.

Red Suns and Red Moons

The picture on page 151 of the Moon rising from behind some trees poses a further question: why is it red?

To start: of course, the only reason we see the Moon at all is that it is being lit up by the Sun, though it is interesting to note that the amount of sunlight that actually reflects off the Moon is not that high. Take an ordinary mirror here on the Earth – it reflects about 95 per cent of the light that hits it. As for things 'out there': Venus, for example, is a shiny thing covered in thick white clouds that reflects about 78 per cent of the incoming sunlight.

For space objects the term *albedo* (from the Latin word 'albus' meaning 'white') is used to describe how reflective something is. The albedo scale goes excitingly from zero (no reflectivity and therefore dark) to one (reflects everything and therefore bright), and is linked to percentage such that 50 per cent is equivalent to an albedo of 0.5, 15 per cent is 0.15, etc. So we say Venus has an albedo of 0.78 – a very high degree of reflectivity indeed.

So, would you care to guess the moon's albedo? Take your time because once you've done this you'll know the answer for ever more. For those who cannot stand the tension, the answer is 0.07. Correct, just 7 per cent of sunlight is reflected off the Moon, making it a very dark object. It's quite strange to think, when you're gazing up at a brilliantly bright Full Moon, that if you were standing on the surface and picked up

some moon 'dust' it would look very much like holding a handful of coal dust here on the Earth – but there you are.

Now, it may only be 7 per cent, but it is 7 per cent of *sunlight*, and light will do exactly the same thing whether it comes directly from the Sun or indirectly off the Moon. Basically I'm saying that the reason for a red Moon is exactly the same as for the red sunset – great, I can cover two things at once.

Imagine, if you will, a tiny packet of light zooming through space. Nothing is in its way until it hits the Moon. With no thought at all it bounces off towards the Earth, and now the only thing in the way before it gets to your eye is the atmosphere. Here's the thing: the combination

The red light of the sunset here has turned not only the clouds red, but also the peaks and snow on the slopes of the Himalayas. (Image courtesy Jason Hughes)

of various gases that surrounds our planet does things to the light. And the amount of air that is in the way affects how and what things are done to the friendly little light packet.

What we need to know is that the light we see is really a part of a whole spectrum of energy. So, for exactly the same reason, we can explain why the sky is blue. Take a look at the diagram below:

Light from the Sun (or Moon) looks as good as white to us, when actually it's a merged mass of varying colours. We always use the usual suspects of red, orange, yellow, green, blue, indigo and violet when we talk about the visible spectrum, but it's all much more subtle than that. Just look at the rainbow, and you'll see not all of the colours are of the same

Visible light – the kind we can see – is just a tiny part of the whole electromagnetic spectrum. The ordinary 'white light' of outside can be revealed as the rainbow of colours it is really made of by using a prism or something similar. You may have some crystal or glass that the sunlight catches as it comes through a window, so you see a rainbow on the floor. The glass is spreading out the merged 'white light' into a spectrum of colours.

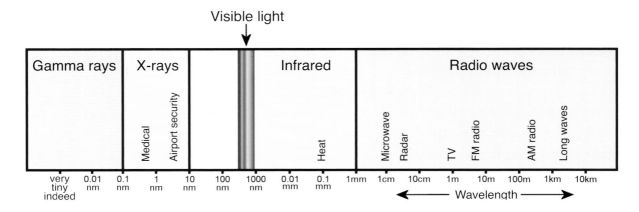

brightness, or width, and there are no definite boundaries between one and the other.

In this part, just think of light as a wave and you'll be fine. These light waves may be spread out – like a gently undulating ocean – or closer together like a choppy sea. The height of each wave doesn't vary, just the distance between them – that's the wavelength. The reason for the colours in a rainbow is that the wavelength of the light as we move, say, from outside the rainbow inwards is getting shorter, and each wavelength has a slightly different colour. So (say the next few words s-l-o-w-l-y), up and down and up and down goes the tender swell of red waves (they have the longest wavelength) while as we move through the rainbow, we find the orange sea is slightly more undulating. Yellow's waves are closer still – you get the idea – down to the shortest blue-end waves. Everyone is on deck and not feeling very well at all by this stage. The reason we see a rainbow is that every wavelength of light reacts differently as it hits the water drop and bounces around inside it before they all emerge on a slightly different path. Red light is affected less than blue light. The bending of light when it moves from one thing through another (like air into a water drop in the case of a rainbow) is known as *refraction*.

You can remember the change in wavelengths and associated colour by memorizing Relaxing Red to Bumpy Blue. You can think of something appropriate for indigo if you wish, though some people think indigo doesn't actually exist. Some scientists think that Isaac Newton, who gave the the seventh colour in the spectrum its name, believed that seven was a universally important number (seven days of the week, seven Western musical notes etc.) and that, even though he only saw six colours, he made one up for completeness. It is true that some humans cannot see this – can you? – which has led to the questioning. I like indigo very much, so I am not with those indigo-less scientists.

So, we've seen that the white (with a hint of yellow) light from the Sun is really a merged spectrum of different wavelengths of light (i.e. colours), and that these react in different ways depending on what they hit or go through. This is the key to blue skies and red sunsets.

Let's take the blue sky: from what I've been saying there is a distinct possibility that you are now thinking that the blueness we see comes from the blue wavelengths within sunlight. And if that is the case you are right: this is exactly where the blue comes from. All the sunlight ploughs into the atmosphere, unaware of the events to follow; for the shorter blue wavelengths

Sunlight being refracted inside raindrops causes the wondrous rainbow phenomenon. The long red wavelengths are refracted less than oranges (not the fruit), which are refracted less than the yellows. The short blue waves are refracted the most. It's due to this changing wavelength with its associated difference in refraction that we see a rainbow at all.

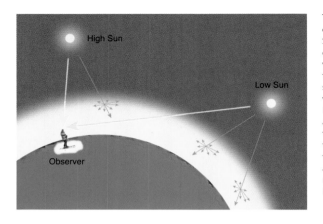

To get to the skiing observer, the high Sun's light, as depicted by the yellow line and arrow, travels a much shorter distance, while the low Sun's light (orange line and arrow) travels much further. The longer the light path through the atmosphere, the more chance it has of being deflected, reflected, scattered and battered about by stuff that is in, and makes up, the air – leading to effects like a red sunset. Of course, sunlight is not just heading my way but everywhere, and the explosions of arrows are where some of the blue light is being scattered out of the 'white' sunlight to give us our super blue sky.

are the right size to interact with the bits that make up the very air itself (oxygen and nitrogen). These interactions actually happen with all colours, but are much more common with the blue-end shorter varieties. The collisions result in a lot of blue light being scattered all over the place – hence the blue sky. The chap who discovered this was Lord Rayleigh, so if you want to impress people the technical term for a blue sky is Rayleigh scattering.

Meanwhile, as mentioned, the longer (red) wavelengths are not affected as much and they race on oblivious to the fate of their blue buddies. Unhindered, the rest of the 'white' sunlight (missing a bit of blue) hits and lights up the ground, trees, penguins, etc.

Of course there is a lot more in the atmosphere than just air. There's pollution, dust, water, sand (from the odd sandstorm), volcanic ash (from the odd volcano) – but we can bunch all these things together and simply call them *particles*. It's these particles that can affect other colours (the longer wavelengths) and the amount of them and where they are in the atmosphere can have an effect on the sky colour, and also the colour of the Sun. This is

where we finally arrive at the first question – the one about the red sunset and red Moon.

When the Sun is high in the sky its light has a relatively small amount of atmosphere to get through before it reaches the ground – a fine helping of Rayleigh scattering for that blue sky – while any particles hanging around do little generally to affect the scene. However, as the Sun gets lower, its light has to travel through more and more atmosphere in order to get to where you are. This greater distance leaves the light open to many more interactions. Almost all of the blue can be scattered out, and if there are plenty of particles in the air (maybe after a prolonged dry period), then colours can be reflected over the sky. This has several consequences: most notably, the Sun becomes redder in colour as more and more of the other wavelengths hit things and, over the longer track, don't get to you. Red light, you see, with its long wavelengths, is very good at ploughing on regardless. Also at sunset, with more of its light scattered by molecules and reflected away by larger particles in the air, the Sun appears dimmer. I would not recommend this as it is extremely dangerous to look too often

I had to go to Waikiki to get this fantastic sunset for you to enjoy. What a warm pleasant evening that was. The sunlight having to travel much further through the atmosphere when the Sun is near the horizon causes the red colours of the sunset. The shorter blue wavelengths of light are filtered out as they scatter off gases and reflect off particles like dust and pollution.

LEFT: Here's a cosmic straw in a glass of water. If you look down the length of the straw it appears warped, bent, broken and of differing size. This is because the light from parts of the straw, on its way to your eye, has to travel through varying amounts of glass, water and air. Each of these has a different 'refractive index', meaning that as the ray of light moves from, say, the water into the glass, there is a slight shift in the angle the light is taking – it has been refracted.
RIGHT: The experiment is finally sabotaged by my daughter!

or too long, but the view of the dim red Sun disappearing over an ocean or sea horizon is truly an amazing spectacle.

And approximately 1,600 words later, this is also why the Moon is red. The sunlight reflected from the Moon behaves in exactly the same way as direct sunlight, so its light too is subject to increased scattering and reflection when near the horizon. All done. Well nearly…

There is another effect that can be added to make a dramatic red Sun. The atmosphere itself refracts things, just like a straw in a glass of water (see the example above).

The thicker the atmosphere the greater the effect, hence things low down have been 'moved' (refracted) upwards from their actual positions. What this means is that when you 'see' the Sun (let's not get into safety again) on the horizon, it

has in fact already set! If you don't believe me, just remove the Earth's atmosphere completely at sunset and you'll see I'm right – the Sun is not there. Magic. Remember that rainbow – look at the outside red band. Now imagine a whole rainbow with only the top red bit peeking out from over the horizon – the rest of it is there, it's just below the horizon (impossible to do, but not to imagine). In essence that's what the Earth's atmosphere has done to the Sun's light. Most of the other light has been refracted out and isn't getting to you, so you're just left with the long red wavelengths. Light does the craziest things!

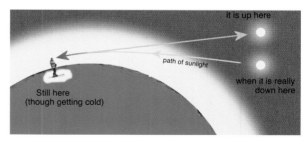

The atmosphere refracts (or bends) light. The longer the light path through the atmosphere, the more it's going to bend. When the Sun seems to be on the horizon it can actually be more than its whole diameter lower, but the light has been refracted round to your position. The final direction that the light hits you from – shown by the red arrow – projected back onto the sky is where the Sun appears to be in the sky – along the orange line. It's very clever, but slightly off-putting, that the Sun sometimes isn't really where you think it is. Oh, yes, and it's the same for the Moon too, and the stars and planets.

CHAPTER 10

Glows and Ghosts

'Look, it's behind you! Over there, lurking between Coma Berenices and Hydra – can you see it?'

Once it's pointed out to you by someone who 'knows', you wonder how you could have missed anything so obvious. What I'm going on about are things such as the **Zodiacal light**, **Zodiacal band** and the **Gegenschein**. These are all somewhat elusive faint glows in the sky – and impossible to see for those living in towns. Even astronomical scribblings that mention them occasionally use words such as 'it has been suggested by some observers that...' and 'there seems to be...' – all of which add to the mystery of these ghostly forms.

What's true is that it is their faintness that makes especially the Gegenschein and zodiacal band so difficult to see.

I'll mention the Gegenschein briefly, as this was my moment of enlightenment, before starting properly with the zodiacal light. There I was, with a few astronomical colleagues, in the desert, a fair distance from Tucson, Arizona, standing with astronomer David Levy in his telescope enclosure. We were there to do a spot of planetary observing in a perfectly dark moonless sky: Mars and Neptune were on the

cards for tonight. It was a warm October evening, and astronomical conversations drifted around. I casually mentioned the elusive Gegenschein and wondered whether I would ever get to see it. David looked up at Aries and said, 'Yep, there it is.'

Surely life wasn't that easy? I peered in the direction suggested, and after confirming that the extremely faint oval glow centred on and orientated with the ecliptic was indeed the Gegenschein, I felt both overwhelmed and slightly bothered. Overwhelmed at finding it, and at the simplicity of finding what I had imagined would be an almost impossible phenomenon. Bothered because I now knew what I was seeing, and *had* seen on many previous occasions before. After years of believing this to be impossible, and worrying about it, I found it staring me in the face. Yes, indeed folks.

I guess it came down to being shown exactly where and what to look for by an expert who knew the sky implicitly. Now I know, I hope the following will enable you to locate these faint hints of misty nothing (am I getting the faintness point across okay?) in the night sky. Someone once commented to me that it was like looking for something that is 'a different shade of black', and you certainly need to spend a good time gazing at the sky at the right time and under the right conditions. But before we go into that, let's look at what the zodiacal light, zodiacal band and Gegenschein are.

Out there in the solar system there are many different things of different sizes: one very big star, a dozen or so objects we could call planets (let's not go into detail here), over 100 planetary moons, minor planets or asteroids, comets and

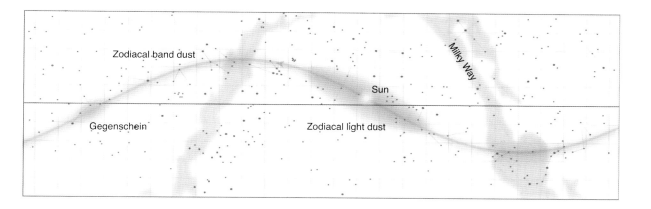

In the image: Zodiacal band dust · Milky Way · Sun · Gegenschein · Zodiacal light dust

dust – tonnes and tonnes of dust. It is the lighting up of this dust that gives us our three phenomena.

Where does this cosmic dust come from? A few places: there is some left over from the creation of the solar system, some from comets, which are big dirty snowballs that release dust as they approach the Sun (in a nutshell anyway), and some from asteroids, or bits of them, that hit one another and throw dust into space.

We know that basically most of these objects orbit the Sun in the same plane – around the ecliptic as seen from the Earth (see Chapter 2 if you don't already know about the ecliptic and stuff like that). Therefore, as the dust comes from objects within this plane, it is to be found all the way along the ecliptic.

Just like car headlamps lighting up mist on the road, the Sun lights up (reflects off) the dust and that's how our various sky glows are created. Don't think, however, that if you were out in space your spacesuit would get covered in dust; each particle that goes to form one of our glows is about 8km away from the next one.

So, time to look at each one: the zodiacal light is by far the easiest of the three to see – at

The Starry Skies map from Chapter 2 has made a welcome return, showing where the dust lies that gets lit by the Sun to produce the zodiacal light, zodiacal band and Gegenschein. There is more dust around the inner solar system, so the brightest section is always close to the Sun – this is the part that creates the zodiacal light. The name zodiacal comes simply from the fact that all the dust is in the plane of the solar system – shown as the ecliptic wavy line across the map – and therefore passes through the constellations of the zodiac. (Ecliptic and stars plotted by Scientific Astronomer software, Wolfram Research, Inc., Mathematica, Version 5.2, Champaign, IL [2005])

times it can match the brightness of the Milky Way (see the top picture overleaf). This is observed as a triangular glow in the west well after dusk, or in the east before dawn, stretching up from where the Sun is (below the horizon) along the ecliptic. The best place to view it is in the tropics and equatorial regions where it appears virtually every clear morning and evening. In fact, in these locations it is known as the 'false dawn', which gives you some idea of how noticeable it is.

The zodiacal light's size at its base along the horizon varies from around 20 to 40 degrees, and decreases to about 5 degrees at its highest point (the top of the triangle), which itself is anywhere up to 50-ish degrees above the horizon. These are not insignificant figures by any means.

Tunisia at 5.10am on 5 October 2005. The Milky Way is on the right and flows up to the top centre of the image, while the zodiacal light is on the left and tilts up to the right. (Image courtesy Dr Francisco Diego/CosmicSky Productions/HLPS)

The best time to look for the zodiacal light:

Location	Best dates	Looking
Mid-northern latitudes (above 40°N)	Mornings – September to November Evenings – February to April	East (pre-dawn) West (post-dusk)
Mid-southern latitudes (below 40°S)	Mornings – April to May Evenings – August to September	East (pre-dawn) West (post-dusk)
Equator and tropics	All year	You've probably got the idea

The zodiacal light can only be well viewed when the ecliptic (blue line) makes a steep angle up from the horizon. Around the equator of our planet the ecliptic is at a high angle all year, hence its ease of viewing there. (Charts and ecliptic created by Scientific Astronomer software, Wolfram Research, Inc., Mathematica, Version 5.2, Champaign, IL [2005])

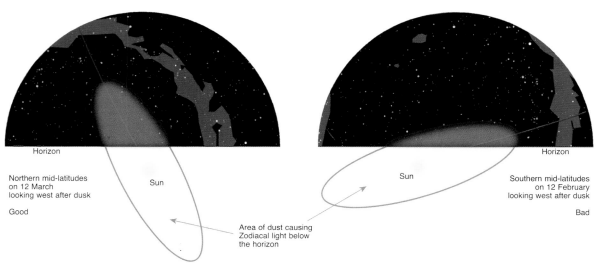

Horizon

Northern mid-latitudes on 12 March looking west after dusk

Good

Sun

Area of dust causing Zodiacal light below the horizon

Sun

Horizon

Southern mid-latitudes on 12 February looking west after dusk

Bad

You may be wondering why there are 'best dates' for certain latitudes, while other places (like the tropics) seem to be good-to-go at any time of the year. It is all down to the angle of the ecliptic to the horizon. The higher the angle the better your chances, as the glow will not be washed out by the thicker atmosphere with its haze, light extinction and/or light pollution when lower down (have a look at the diagram bottom left to see what I mean).

Here's a picture of the zodiacal light taken by the astronauts on the Apollo 15 command module *Endeavor* just before sunrise on 1 August 1971 as they orbited the Moon. (Image courtesy NASA)

Well, that's the bright one done – we now get into the ghostly realms of the zodiacal band and the Gegenschein.

Because most of this dust hangs about (orbits around) in the inner solar system and tapers off as it gets out towards the asteroid belt, when we look away from the Sun we're obviously looking into a less dusty area – but it is all one cloud. So the top of the zodiacal light tapers off, merges imperceptibly and becomes the zodiacal band, which then continues to follow the ecliptic across the sky.

Now, well after sunset, with the ecliptic high in the night sky, if you follow the zodiacal band to the point directly opposite the Sun, there will appear an oval-shaped, marginally brighter larger patch of the zodiacal band: the Gegenschein – this is a German word meaning 'counter glow'. It is around 10 degrees long by 6 degrees wide, although this is difficult to pin down accurately due to the fact that it is so faint.

The reason the Gegenschein is slightly brighter than the zodiacal band is the same as the reason the Full Moon is brighter – the Moon is on the opposite side of the sky to the Sun, so we see the entire lighted side and no shadows. Think of those tiny dusty bits as tiny moons, all lit in the same way. Just like the Moon, once they move away from the point opposite the Sun, we start to see shadows and not the whole of the lighted side (hence the phases).

There are two factors that explain why the Gegenschein and zodiacal band are so faint: one, there is less dust to light up in the direction away from the Sun, and two, there's the greater distance of the dust from the Sun itself to consider – here the *inverse-square law of light* comes into play, meaning the further the light has to travel from the source to the thing it's lighting up and then on to the observer (you), the fainter the thing is.

With this faintness in mind, any of the following can remove your chances of observing the zodiacal band and Gegenschein: light pollution, moonlight, nearby bright planets and the Milky Way. The Milky Way? Indeed, the overpowering light from this band in a truly dark and clear sky will not allow your eyes to dark-adapt enough to see our two extremely faint glows. The Milky Way is also much brighter than the glows, so if it is running over the Gegenschein, you've had it. Needless to say, if you cannot see the Milky Way from where you live due to light pollution, you have no chance anyway.

The same brightness issue happens with planets. When I saw the Gegenschein, Mars was shining next door at magnitude -2.2. Typical! However, shielding the planet with my hand, I located the glow without too much trouble.

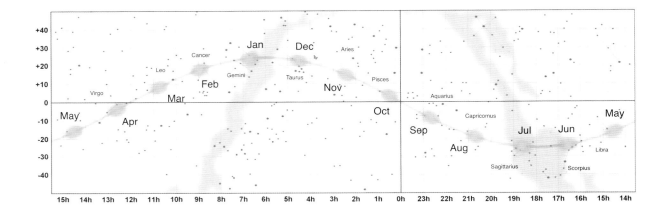

So you really want to look for the Gegenschein? Well, here's the chart of where the ghostly glow resides on the first of each month. Remembering that it is washed out by the Milky Way, there is no chance of finding it from mid-May until the end of July when it hides over (or very close anyway) the bright heart of our Galaxy in the direction of Scorpius and Sagittarius. Similarly from mid-December until mid-January it's in front of the Milky Way in the Taurus and Gemini region. (Ecliptic and stars plotted by Scientific Astronomer software, Wolfram Research, Inc., Mathematica, Version 5.2, Champaign, IL [2005])

The Gegenschein charts above and opposite show where it will be throughout the year. Obviously there are times when viewing will be impossible, but away from those, if it's dark, have a go. Look slowly around the area in question, taking your time, and try to notice any even vague variation in the 'brightness' of the black – I know, but that's how it is. Even if you get a hint that there is something there, you may well have located your glow. Don't fool yourself, though; the mind is very good at playing tricks on us.

Don't take 'too much trouble' to mean it was that easy. I was shown exactly where to look, and I had many years of experience of gazing into the night, so I was used to the 'shades of black' mentioned earlier. Your best chances of seeing it and the zodiacal band are when your eyes are fully dark-adapted – maybe after an hour outside – and the glows are as high in the sky as possible.

However, once you are satisfied you have found the genuine articles, you can tick a few more boxes of Solar System objects that not many astronomers have seen. Good luck.

The best time to look for the Gegenschein:

Location	Best dates	Looking in
Mid-northern latitudes (above 40°N)	February to mid-March October and November	Cancer and Leo Pisces and Aries
Mid-southern latitudes (below 40°S)	April to beginning-May August and September	Virgo and Libra Capricornus and Aquarius
Equator and Tropics	February to beginning-May August to December	Cancer to Libra Capricornus to Taurus

As with the zodiacal light table on page 160, the equator and tropics come off the best with the Gegenschein. This is not specifically due to the high angle of the ecliptic this time, but simply because this section of the world always carries the most interesting parts of the sky really high in the sky. In the mid-latitudes we have to wait our turn, as there is no point in trying to look for something faint low down in a gunky sky.

For those in need of proper coordinates, here's where the Gegenschein is centred during the year:

Date – first of the month	RA	dec.	Constellation (on first of the month)
January	6h 45m	+23° 02′	Gemini
February	8h 58m	+17° 11′	Cancer
March	10h 47m	+7° 43′	Leo
April	12h 41m	-4° 25′	Virgo
May	14h 32m	-14° 59′	Libra
June	16h 35m	-22° 00′	Ophiuchus
July	18h 40m	-23° 08′	Sagittarius
August	20h 44m	-18° 06′	Capricornus
September	22h 40m	-8° 24′	Aquarius
October	0h 28m	+3° 04′	Pisces
November	2h 24m	+14° 19′	Aries
December	4h 28m	+21° 45′	Taurus

Aurora

Aurora over Dundee, Scotland. (Image courtesy Harry Ford)

With meandering curtains of reds and greens revealing the magnetic forces at play through our atmosphere, a large shimmering aurora must have a good chance of winning the award for the greatest night-time spectacle – maybe equalled by a big bright comet or a truly magnificent meteor display (but these can be hundreds of years apart).

Anciently speaking, Aurora is the Roman goddess of dawn – a beautiful woman who flies around in the early mornings, sometimes on the horse Pegasus, sometimes being pulled along in a chariot or sometimes flying solo using her wings. Whatever, she's there to announce the arrival of her brother, the rising Sun. On good days she also pours the morning dew from a very big jug.

The practical naming-upshot of this is that those wondrous lights seen in the far northern hemisphere known as **aurora borealis** translate as Northern Dawn, while **aurora australis** is the Southern Dawn for those deep down in the southern hemisphere. Alternatively we know them as the northern or southern lights, or even just as the polar lights. Indeed 'polar' really gives an indication of where aurorae are made and seen – that is around the poles of the Earth, or more correctly, the magnetic poles.

Because our planet is just like a big giant magnet in space, it has associated big giant invisible magnetic fields looping around and radiating out into space. The simple experiment with an ordinary bar magnet under a sheet of paper onto which you shake and jiggle iron filings reveals the same kind of invisible magnetic field lines on a much smaller scale.

Now, as hinted at above, the magnetic poles of the Earth are not lined up with the spinning axis

Looks like just another ordinary night for Dick Hutchinson, up in the small town of Circle, Alaska. The town's founders thought they were on the Arctic Circle, hence the name, but the risings and settings of the Sun would have informed them otherwise (if they had bothered to take any notice) – it's 77km south, in fact. Maybe they were awestruck by the incredible aurorae they saw and decided to stay put. Who would blame them? Dick Hutchinson, the photographer of these amazing shots, clearly thought the same – the top-left image shows the curtain of lights weaving its way over his house. He is indeed a lucky guy. Commenting on the lower-left car, snow and bluish aurora picture, Dick said, 'I never got far from the pickup as it was -47°F (-44°C) at the time.' (Images courtesy Dick Hutchinson)

(what we know as the North and South Poles), but offset to them by about 11 degrees. Actually, it is not possible to pinpoint exactly where the magnetic poles are at any given time, as they can wander around by tens of kilometres a day. In practical terms this means your compass does not point truly north, but to wherever magnetic north happens to be – they are fairly close but adjustments have to be made for important Earthly applications like navigation.

Aurorae are made when charged particles thrown out by the Sun cause an overloading and subsequent instability of the Earth's magnetic field, leading the particles to cascade down the invisible magnetic lines towards the poles. There they smash into the atmospheric gases, which go a bit crazy and then get rid of all their excitement by producing some light. Of course, the more particles that do this, the more the light that's made, and with any luck there's enough to build an aurora.

The Sun really is the key here: there is a clear link between the number and size of sunspots (which have intense magnetic fields associated with them) and auroral activity down here on Earth. Plus, once a week – more when the Sun is particularly busy – it releases a great bubble of gas wrapped in solar magnetic field lines in an event known as a *coronal mass ejection*. These can cause disruptions to satellites in orbit and to power supplies on Earth, but they can produce some incredible auroral displays.

Where it's happening and what it looks like

This all goes on at heights of around 80-200km, but aurorae can reach up to 300km or more above Earth's surface. The marvellous colours produced depend on the energy the particles have when they hit certain types of gas – above 200km reds are caused by oxygen, between 100 and 200km nitrogen glows blue and this in turn causes excited oxygen to produce the main green colour of an aurora; below 100km nitrogen emits a deep reddish colour.

While that's going on, the flowing magnetic lines mean that the 'lights' can take different forms during a display that can last many hours. A low, arching, gentle green glow to the north is generally the first sign of an aurora. Vertical red and green *rays* emanating from the arch as it grows and ripples into a long *arc* are joined by friends that stretch like *ribbons* right across the sky – sometimes taking on the appearance of *curtains* blowing in the breeze.

Somewhere in Aberdeenshire, Scotland, sometime in November 2004, about midnight, this picture was taken.

Here at around 57 degrees north you can expect to see an aurora a couple of times a year, but it really depends on

what the Sun is doing. (Image courtesy: Dr Francisco Diego/CosmicSky Productions/HLPS)

Then there is the *auroral corona*, with its rays appearing to converge right above your head – just an effect of perspective.

These magnetic atmospheric phenomena are easier to see in the northern hemisphere purely because this is where there is more inhabited land. Take a look at the North and South Pole aurora activity plots below. In the southern hemisphere the activity (the round orangey doughnut thing) sits over Antarctica, and there's a lot of water before you get to any other land where you have a chance of seeing it – Tasmania, New Zealand, southern Chile and southern Argentina. 'Up north' it's Iceland, Norway, Sweden, Finland, Russia, Canada, Alaska and Greenland that are all much closer to any activity.

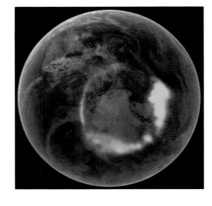

Here's the ring of aurora circling Antarctica as seen from the IMAGE satellite on 11 September 2005. (Image courtesy NASA)

This is not to say that these are the only places you can go: during increased solar activity aurorae have been known to reach all the way to the equator, although this is extremely rare.

As for viewing opportunities, of course you need dark skies, which means September to April for aurora borealis and March to October for aurora australis. The only problem is, as many have found, there is no guarantee, even if you are technically in the right place at the right time, that you'll see one. Many have ventured north only to have to wait for weeks before an aurora appears.

The particles flowing down the magnetic field lines of the Earth hit the atmosphere in a circle-ish shape called the auroral oval, which is centred on the Earth's magnetic pole. On the plots the centre of the red dotted lines are the North or South Poles, but you can see that the auroral ring is not centred on this, as discussed above. The entire ring is around 3,000km in diameter (but can grow larger when there's lots of solar activity) and normally sits between latitudes 60 and 70 degrees north and south.

The colour bar to the right of the plots is in the legendary 'ergs/cm²/sec' atmospheric energy input unit (means nothing to me, actually, but

On the left is a view above the North Pole, while the right shows what's going on at the same time over the South Pole. The orangey circle is the important thing – this is where all the activity is happening. The outer bluish areas just show the extent of the 810km-high polar-orbiting NOAA-17 satellite's data. (Image courtesy Space Environment Center, Boulder, Colorado/National Oceanic and Atmospheric Administration US Dept. of Commerce. Phew!)

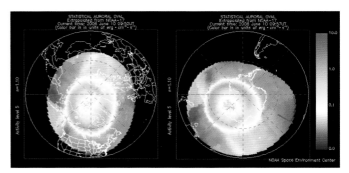

the following is a breeze to understand) and I have the great Dave Evans at the Space Environment Centre (SEC) of the National Oceanic and Atmospheric Administration (NOAA) to thank for these references:

Here are the registered energy levels on the scale and what that gives us in the sky (or not, as that case may be). From now on whenever you get aurora forecast plots you'll know what's what:

0.1 erg/cm²/sec (white on the plots) would produce a sub-visual aurora – that is, we couldn't see it, but instruments could pick up the light.

0.5 erg/ cm²/sec (yellow on the plots) produces an aurora that a person with dark-adapted vision under good seeing conditions (no moon, no high cloud, no city lights) can just about see.

5 ergs/ cm²/sec (orange on the plots) allows a person to see the aurora pretty quickly after going outside from a lit room.

50 ergs/ cm²/sec (very red) means you just have to look out the window and there it is.

Very rarely the counter reaches 500 ergs/ cm²/sec, which means you could read a newspaper by the auroral light being produced – as if you'd be reading a newspaper!

These images show that being in orbit is one of the best vantage places for observing aurorae. Here are two wonderful shots of aurora australis as seen by the astronauts on the shuttle mission

STS-39 back in 1991. As their orbit was 260km in altitude they were actually flying below the spiky red tops of the right-hand aurora image. (Images courtesy NASA)

A Final View

The top of the La Palma island volcano (one of the Islas Canarias) is home to some fine observatories, due to the clear, dark skies to be found there. About 20 years ago, this was where I, and a group of fellow astronomers, viewed one of the most stunning night skies we've ever seen. The entire scene was literally breathtaking in its magnificence. The Milky Way flowed overhead, but, to be honest, the stars were sparkling absolutely everywhere and shining brightly right down to the crisp, non-hazy horizon – the rotating Earth popped each star in or out of view, depending on the direction. It was just superb.

Therefore, naïvely, I thought that all mountains with great telescopes on, especially the one I'm about to mention, would provide me with that same experience of the sky. It took me over 15 years to discover that that was not really the case. This time I stood on the volcano Mauna Kea, in Hawaii, gazing upwards while standing next to some of the finest observatories in the world. 'Come on, sky, show me the Universe.'

Oh. Nope, not as good as La Palma. This wasn't how it was supposed to be. The skies were amazing, yes, but not stunning. I startled myself with this thought. After returning to Waikiki (well, someone has to), I pondered for a while. Why was the view not as good? On both the trips – to La Palma and Hawaii – the view of the sky was one thing I was really looking forward to, so I always had 'anticipation' in my mind.

In hindsight, hoping for something that you remember well from the distant past to repeat itself is always a bad move. You're just asking for trouble. Having said that, at least being able to recall such experiences always brings a smile. I once said to one professional astronomer that I think appreciating the wonder of the night sky – just gazing upwards into the starry abyss, speculating on alien life, or black holes, or whatever – is like appreciating a sunset. Both involve us being witness to and taking part in some wonderful aspect of nature: whether taking an artistic or scientific approach, or mixing them all up in a bag and giving it a good shake. 'They' laughed somewhat at this ridiculous idea, and, to be honest, I was glad 'they' weren't with us on La Palma. Not everyone looks at sunsets the way I do, and probably not the starry realm.

It seems a great shame that some people will never appreciate the beauty of the heavens simply with the unaided eye.

Maybe I was remembering La Palma too fondly? Yes, that is probably true. Then I realized that it is not just the sky itself that makes the experience. It's who you are with, what they are interested in, what you've been doing all day and all manner of factors that affect what and how you see something and how you feel about it.

La Palma had been a holiday with great friends, while Hawaii was work, and it being over 3,500m above sea level meant that the work was particularly tiring. I'm not feeling as much sympathy from you as I would like about this, but anyway, this led to two fundamentally different mental and physical approaches to my views of the Canary and the Mauna Kean night skies.

Further, thinking about 'great skies I have seen', they always involved times of quiet observing, friendly starry banter outside during a party or group astrocamps. And this can be anywhere from the darkest location to a light-polluted town – they all have something different and splendid to offer if you are in the right frame of mind. Therefore, wherever you are, the night sky has something to give, but you'll almost never experience identical skies identically the same way.

Starry Skies,

Astroglossary

Astronomical Unit (A.U.)

One of these units represents the mean distance from the Earth to the Sun. It is mainly used within solar systems (ours or alien ones) as a way of explaining things that are a long way apart without using awkward large numbers. For example your alternatives to the lovely 1 AU are: 149,597,871km or 92,955,807 miles or 81,801,110,564 fathoms – all of which amazingly equates to 299,195,742,000 Cambodian hats. Isn't the internet wonderful?

Aurora

This is a night-time display of various coloured glows and shapes, seen mainly in the higher latitudes in both hemispheres, in a ring around the magnetic poles. Aurorae are caused by little charged particles from the Sun that are channelled by the Earth's magnetic field down to collide with and excite the particles that make up our atmosphere. It's the release of this general excitement that produces the lights we see.

Black hole

You can make one by piling loads of old jumpers (or stuff) together until the escape velocity needed to get away is greater than light. Supermassive black holes like these power the centres of many galaxies. Another type is the remains of a massive old dead star. Fuel, in the form of gas, keeps a stellar furnace shining brightly, but once that fuel runs out the pressure holding the star up stops and the force of gravity takes over – it collapses in on itself. This enormous 'sucking' force of gravity becomes so strong that not even light can escape and it 'appears' black, hence the name. I've got one of these in a box in the loft, but I'll only use it in an emergency.

Brown dwarf

There's a murky world between planets and stars, and this is where brown dwarfs live. Large planets (much bigger than Jupiter) can start to have some nuclear reactions in their core, but they are not large enough (and therefore the temperature isn't high enough) for the reactions to be the hydrogen-burning variety that you find in normal stars. So they only produce a small amount of heat which can be picked up in the infra-red (heat) region of the spectrum only.

Day

The apparent **sidereal day** is the time taken for the Earth to spin on its axis once with respect to the 'fixed' background stars – of course this means an exact 360-degree spin of the planet, which takes 23 hours 56 minutes 4 seconds. Compare this with the apparent **synodic day**, which is the length of time it takes a planet to spin once in relation to the thing it's orbiting. For the Earth this thing is the Sun and leads to this period also being known as a **solar day** – our usual 24-hour day representing an Earth spin of about 361 degrees. More about this in Chapter 1.

Declination & right ascension

These two terms refer to the starry-sky mapping coordinate system, which is not as complicated as it sounds. Basically declination (dec.) and right ascension (RA) serve the same purpose as latitude and longitude on Earth – they enable you to pinpoint where something is in the skies. Dec. mirrors the lines of latitude, even to the extent of being measured in degrees, while RA is like longitude, but uses hours instead of degrees. Don't worry, the conversion is a piece of cake: 1 hour RA equals 15 degrees, therefore 24 hours equals 360 degrees, which is once all the way round the sky. Both scales start from zero – declination goes north or south from the celestial equator and, as with latitudes on Earth, the numbers get bigger the further away from the equator you get; Right Ascension begins at the March equinox point and goes round in a circle from there. Look back at the Starry Skies map on page 12 if you have trouble visualizing this.

Ecliptic

From a Greek word meaning 'place of eclipses', this imaginary line is the path that the Sun takes around the sky over the year. It passes through twelve constellations, giving rise to the 'signs of the **zodiac**'. The word *zodiac* itself means 'line of animals', and our word *zoo* comes from the same source. Because the solar system was all made from a disc of space stuff, we find that the Moon and planets also stay close to this line.

Elliptical orbit

An orbit around something that is an oval shape instead of being circular. The more elliptical the orbit, the more squashed the oval. The Earth has a slightly elliptical orbit around the Sun, which means that in January we are 5 million kilometres closer to the Sun than we are in July.

Equinox

From the Latin for 'equal night', for it is the time when most places in the world experience day and night of approximately equal length. At this one instant neither the North nor the South pole has any 'tilt' towards or away from the Sun. There are two equinoxes, around 21 March and around 23 September, the former being used to define the zero point for right ascension (see **declination & right ascension**, above). The time when one or other of the poles is tilting as much as it can towards or away from the Sun is called the **solstice** (see below).

Gegenschein

An extremely faint patch of light seen in very dark (did you read 'extremely' and 'very' as seriously as it was meant?) skies. German for 'counterglow', it sits on the ecliptic exactly 180 degrees from the Sun – which means it's at its highest at midnight. This light is due to the scattering of light from tiny dust particles in the plane of the solar system and it's one of the 'glows and ghosts' that Chapter 10 is all about.

Libration

An effect that allows us to see 59 per cent of the Moon's surface in total – though only 50 per cent at any one time. It is due to several things, like the Moon's orbit being tilted and slightly elliptical around the Earth as well as the change in our viewing position throughout the day.

Light-year (ly)

The distance that light travels in one year, used as the standard distance measurement for deep space (there's another unit called the *parsec*, but that is only used by people you never really want to speak to at a party). A light-year is 9,460,530,000,000 kilometres km (9.4 trillion, for short) or 5,880,000,000,000 miles (5.8 trillion).

Magnitude

How bright something like a star appears in the sky is called its visual or **apparent magnitude**. However, this gives no indication of how bright it really is, as apparent magnitude doesn't take distance into account. That's the job of **absolute magnitude**, which mathematically lines all the stars up at a distance of 32.6 light-years so that they can be directly compared with one another. There's loads more about magnitude in Chapter 4.

Month

The **sidereal month** is the time the Moon takes to orbit the Earth relative to the starry background. This time is 27.3 days. The **synodic month** is the time the Moon takes to orbit the Earth relative to the Sun. This is 29.5 days, and is the time during which Moon phases repeat themselves so that two successive Full Moons, for example, are 29.5 days apart. I talk quite a lot about this in Chapter 8.

Nebula

A cloud of dust and/or gas merrily floating around in space. Nebulae can be fairly wispy, like mist, or more dense, akin to a heavy fog. One can usually be made visible by putting a star behind, in front or even inside it. Indeed some varieties can be a birthplace for stars themselves. The box on pp.16–17 gives more details of different types of nebula.

Occultation

This is the time when one object blocks the light from another object, such as the Moon moving in front of Jupiter. If a star or planet just catches the edge of the Moon, then this is known as a *grazing lunar occultation*.

Parallax

A way of finding out how far away things are using positions of the Earth's orbit around the Sun at different times to detect tiny movements in nearby stars. Just like holding your finger up at arm's length, then seeing how the background changes as you look at your finger using your left and then your right eye. You see a shift in the position of your finger against the background, whereas astronomers detect the shift in the position of a nearby star against the more distant background ones. With this shift in hand it's a simple calculation to work out the distance. There's more about this in chapter 4.

Proper motion

All stars have their own individual motions through the heavens, and they give their motions away by slowly changing their positions over time. Even though, relative to the Sun, these motions can be hundreds of kilometres per second, few of them are noticeable over a lifetime purely because the stars are so far away. This extremely tiny movement known as the star's *proper motion* is given in arc seconds per year. I chat about this in Chapter 6.

Right ascension

See **declination**, above.

Solar day

See **day**, above.

Sidereal day/month

See **day/month**, above.

Solstice

This is Latin for 'Sun stand still', for it is the day when the positions of the rising or setting Sun on the horizon momentarily stop moving in one direction and begin to move in the other. For example, in the northern hemisphere, each day for most of December the rising Sun edges slowly to the south until the December or winter solstice (usually 21 or 22 Dec), when the furthest south sunrise is reached. From then on the sunrise moves back further and further north until the June or summer solstice standstill (usually 20 or 21 June) and then back southward the sunrises go. In the southern hemisphere, of course, it is the other way round, with the winter solstice in June and the summer one in December. At the December solstice the South Pole of the Earth is tilted as far as it can go towards the Sun, while six months later, at the June solstice, it's the north pole that leans at its maximum sunward. See also **equinox**, above.

Supernova

A star that ends its days with a massive explosion. A supernova is a variable star to end all variable stars – a one-use-only variety. The tremendous brightening is caused by a final gravitational collapse and sensational rebound. Some are very bright indeed, as was the case with the one that was visible during broad daylight for two weeks in November 1752.

Synodic day/month

See **day/month**, above.

Terminator

The boundary between the day time and the night time of a moon or a planet. In other words, it's the line between what is sunlit and what is not. The Moon is a good example, as it's the only one we can see with the unaided eye: if you were standing on the Moon's terminator you would see the Sun slowly rising or setting. There's more about this in Chapter 8.

Waning moon

The phases of the Moon from Full through to New. The waning Moon first becomes visible in its gibbous phase a short time after the Sun has set, but on the opposite side of the sky. Day by day the Moon moves towards the morning Sun, shrinking past half to crescent, and finally disappears as the New Moon.

Waxing moon

The phases of the Moon from New through to Full. It's this waxing stage that most of us see. The Moon first becomes visible as a thin crescent in the evening sky close to the setting Sun. As the days go by the Moon moves away from the Sun and grows to a half, gibbous, then Full Moon.

Zodiac

See **ecliptic**, above.

Zodiacal band

See **Zodiacal light**, below.

Zodiacal light

A cone of light stretching up before the rising or after the setting Sun, best seen in light-pollution-free locations and where/when the ecliptic hits the horizon at a good high angle. It is caused by the scattering of sunlight from the myriad of tiny particles that live in the inner solar system. There are actually particles orbiting further out, but their numbers rapidly dwindle (I have never used the word 'dwindle' before) as you get beyond Mars. Nevertheless this does lead to a **zodiacal band** stretching around the entire ecliptic, with the oval ghostly **Gegenschein** at the anti-solar point. Boo!

Index

Page numbers in *italics* refer to illustrations.

Acknowledgements

Special thanks to Francisco Diego, Dick Hutchinson, Robert Massey, 'The' Alan Longstaff, Greg Smye-Rumsby, Monika Fischer, Jason Hughes, Joy Costa and Paul Wootton.